7th International Conference on Green Materials and Environmental Engineering (GMEE2022)

E3S Web of Conferences Volume 341 (2022)

Changsha, China
16 – 17 September 2022

Editors:

Mihaela Popescu
Dariusz Jacek Jakóbczak

ISBN: 978-1-7138-4386-3

Printed from e-media with permission by:

Curran Associates, Inc.
57 Morehouse Lane
Red Hook, NY 12571

Some format issues inherent in the e-media version may also appear in this print version.

For additional information, please contact EDP Sciences – Web of Conferences at the address below.

EDP Sciences – Web of Conferences
17, Avenue du Hoggar
Parc d'Activité de Courtabœuf
BP 112
F-91944 Les Ulis Cedex A
France

Phone: +33 (0) 1 69 18 75 75
Fax: +33 (0) 1 69 28 84 91

contact-edps@webofconferences.org

Additional copies of this publication are available from:

Curran Associates, Inc.
57 Morehouse Lane
Red Hook, NY 12571 USA
Phone: 845-758-0400
Fax: 845-758-2633
Email: curran@proceedings.com
Web: www.proceedings.com

TABLE OF CONTENTS

Preface

Great honor to have your supports to 2022 7th International Conference on Green Materials and Environmental Engineering (GMEE2022) and participation in this academic discussion mainly direct at environmental properties of materials, sustainable development of materials environmental protection and other related areas.

GMEE2022 mainly focus on advanced academic researches and studies in these fields including Green Materials and Environmental Engineering, and aims to bring together researchers and practitioners from academia and industry to highlight the importance of computing technologies and data science as well as establish new collaborations in these areas. The convening of GMEE2022 is of great and positive significance to relevant academic research.

And due to current epidemic development and prevention and control policies, this conference was held in virtual form via Tencent Meeting on January 17, 2022. GMEE2022 covered Opening ceremony online, Keynote Speech, Oral Presentations and Poster Presentations from scientists, engineers working in similar areas to establish platform for collaborative research projects in these fields. The keynote speeches included: 1. "Shape Reversibility and Temperature Deformation Relation in Memory Behavior of Shape Memory Alloys" of Prof. Osman Adiguzel from Firat University, Turkey, 2. "Dark Data representation using Turiyam Set" of Dr. Prem Kumar Singh from Gandhi Institute of Technology and Management, India and 3. "Experimental Investigation on AA7075/B4C/ZrC Hybrid Nanocomposite Fabricated by Powder Metallurgy Techniques" of Prof. Krishna Murari Pandey from National Institute of Technology Silchar, India.

We distributed all submissions for peer reviewing based on main topics and experts' research fields. We reiterate that the conference firmly resists the plagiarism. Any act of plagiarism is unacceptable. After reviewing and manuscript similarity detecting, there are 29 papers were selected in proceedings.

At last, we take this opportunity to thank all the authors and teams that have actively participated in GMEE2022 conference, highly responsible reviewer experts in Technical Program Committee, experienced editors, authors and scholars enthusiastic in research. Sincerely thank you for your warm supports and look forward for collaboration with you in the future.

GMEE2022 Organization Committees

Statement of Peer review

In submitting conference proceedings to *Web of Conferences*, the editors of the proceedings certify to the Publisher that

1. They adhere to its **Policy on Publishing Integrity** in order to safeguard good scientific practice in publishing.
2. All articles have been subjected to peer review administered by the proceedings editors.
3. Reviews have been conducted by expert referees, who have been requested to provide unbiased and constructive comments aimed, whenever possible, at improving the work.
4. Proceedings editors have taken all reasonable steps to ensure the quality of the materials they publish and their decision to accept or reject a paper for publication has been based only on the merits of the work and the relevance to the journal.

Title, date and place of the conference

2022 7th International Conference on Green Materials and Environmental Engineering (GMEE2022)
January 16-17, 2022
Changsha, China,

Proceedings editor(s):

Prof. Mihaela Popescu, University of Craiova, Romania
Prof. Dariusz Jacek Jakóbczak, Koszalin University of Technology, Poland

Date and editor's signature

Mihaela Popescu
Dariusz Jacek Jakóbczak

Preparation of ZnO-GO nanocomposites and their properties

Caifeng Jin[1], *Yuqing* Cheng[1], *Wenbin* Liu[2], and *Yuguang* Lv[1,*]

[1] College of Pharmacy, Jiamusi University, Jiamusi, China
[2] School of materials science and Engineering, Jiamusi University, Jiamusi, China

Abstract. ZnO nanoparticles were prepared by hydrothermal method. The modified Hummer process was then used to prepare go. The ZnO-GO com posites were then mixed to produce different ratios of ZnO-GO, characteris ed and their photocatalytic properties were investigated using rhodamine B as the target degradation species. The results showed that the degradation r ates of rhodamine B were 41.96% and 84.42% for the nano-ZnO and ZnO-GO 7% composites, respectively, when illuminated with xenon light for 36 0 min. The improved photocatalytic activity of composite materials is prim arily due to the introduction of appropriate amount of GO, which could ra pidly capture and transfer to the photogenerated electrons, and could load more ZnO and prevent the agglomeration of ZnO particles, increasing the s pecific surface area of ZnO and its activation point, as well as increasing th e adsorption of pollutants on the ZnO surface and improving the photocatal ytic efficiency.

Keywords: Namo-ZnO; Graphene oxide; Photocatalysis.

1 Introduction

As society develops and environmental problems such as waste pollution become increasingly prominent, a sewage treatment method is urgently needed. Semiconductor oxide photocatalytic oxidation technology is a relatively new type of technology for treating water pollution. As a photocatalyst, semiconductor oxides include TiO_2, ZnO, WO_3, SnO_2, etc. Among them, the redox capacity of titanium dioxide is very strong, chemically stable, non-toxic, and does not cause secondary pollution.Ban bandwidth, ban bandwidth zinc oxide closer to titanium dioxide, and zinc oxide synthesis with a low cost, is a metal oxide semiconductor material that can replace titanium dioxide for the effective degradation of the organic dye contaminant [1-2]. Nanoparticles will exhibit many excellent properties compared to normal ZnO, such as non-toxicity, piezoelectricity, fluorescence, and the ability to absorb and scatter ultraviolet light. The main ways to improve the photocatalytic activity of ZnO nanomaterials are: firstly, to control the ZnO growth structure; and secondly, to carry out surface modifications. At present, the problems encountered in the process of overcoming composite graphene are often solved by using surface modification, i.e. replacing graphene with graphene oxide. In this study, ZnO nanoparticles were firstly prepared by hydrothermal method; then graphene oxide (GO) was prepared by modified

*Corresponding author: yuguanglv@163.com

Hummers method; the two were then mixed to prepare ZnO-GO composites in different ratios, and their photocatalytic properties were investigated using rhodamine B as the target degradant[3].

2 Catalyst material preparation

2.1 Preparation of ZnO-GO composites

The ZnO dispersion was obtained by adding 50 ml of anhydrous ethanol to a beaker and sonicating for 2 h. The graphene oxide ethanol suspension was added to the ZnO dispersion (50 ml of anhydrous ethanol was added to GO and sonicated for 2 h to obtain an ethanol suspension of graphene oxide) and stirred at elevated temperature for 24 h. The resulting product was washed by centrifugation and anhydrous ethanol. The product was then re-dispersed in 10 mL of anhydrous ethanol. Vacuum drying for 24 h to obtain ZnO-GO nanoparticles (ZnO-GO1%, ZnO-GO3%, ZnO-GO5%, ZnO-GO7%, ZnO-GO10% were produced respectively)

3 Results and Discussion

3.1 Structural and morphological characterisation

Figure 1(a) shows the XRD pattern of the morphology of the ZnO nanoparticles prepared by the hydrothermal method. All diffraction peaks match the standard data for the fibrillated ZnO structure of the ZnO hexagonal crystal system (JPCDS 36-1451). The sharp shape of the diffraction peaks in the figure indicates that the obtained ZnO crystals are well crystalline. The XRD patterns of both are highly compatible and the absence of their characteristic diffraction peaks in the XRD patterns indicates that the synthesised samples are both of high purity and no impurities are produced. No diffraction peaks of graphene were observed in the XRD patterns due to the small proportion of graphene oxide in the catalyst.The XRD patterns of the composites (c) show that the doping of different amounts of graphene oxide resulted in lower intensity and wider peak shape of ZnO compared to pure ZnO, and a decrease in crystallinity of the composite samples. The effect on the crystallinity of ZnO gradually increases as the amount of graphene oxide doping increases[5]. In addition, the SEM and TEM photographs given in Figure 2 show that the ZnO particles are attached to the surface layer of graphene oxide, and the graphene oxide doping obviously prevents the agglomeration of ZnO nanoparticles.

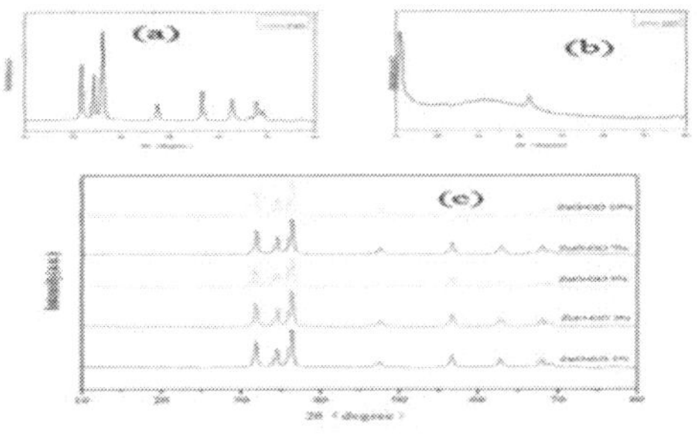

Fig. 1. XRD spectra of ZnO (a), GO (b), ZnO-GO composites (c).

Fig. 2. Scanning electron micro-graphs of ZnO nanoparticles (a) and ZnO-GO1%(b),GO(c), ZnO-GO1% (d) transmission electron micrographs.

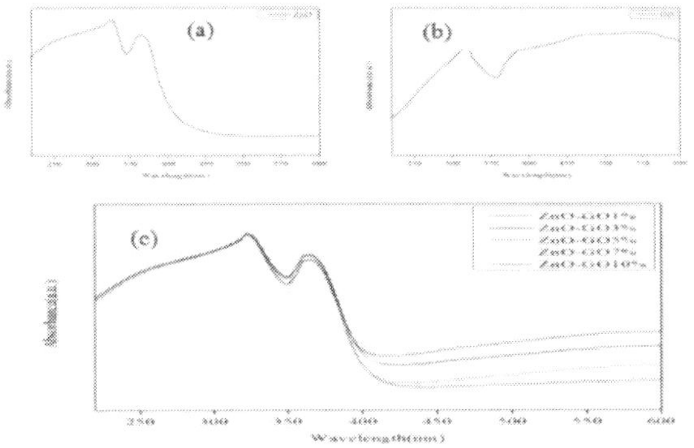

Fig. 3. UV-Vis spectra of ZnO (A), GO (B), and ZnO-GO composites with different concentrations (C).

Figure 3 shows the UV-Vis absorption spectra of different amounts of graphene oxide doped ZnO, from which it can be seen that all samples have strong absorption at 385 nm, corresponding to the intrinsic band width of ZnO nanoparticles. The absorption intensity of the graphene oxide doped photocatalysts was significantly enhanced in the visible range, indicating that the addition of graphene oxide altered the light absorption properties of ZnO.

4 Degradation of rhodamine B by ZnO-GO composite photocatalyst

4.1 Photocatalytic methods

50 mg of photocatalyst and 100 mL of rhodamine B solution at a concentration of 10 mg/L were weighed separately in a 100 mL double beaker. The solution was magnetically stirred in the dark for a certain period. A xenon light source and condensation unit were then switched on for visible light-catalyzed degradation. Approximately 5 mL of the reaction solution is extracted at regular intervals, the catalyst is removed by centrifugation and the supernatant is measured by UV-Vis spectrophotometer at a maximum absorption wavelength of 554 nm to determine the residual concentration of the catalytic degradation and to calculate the degradation rate η%[6]. The formula was calculated as:

$$\eta\% = 1 - C_t/C_0 \times 100\% = 1 - A_t/A_0 \times 100\% \quad (1)$$

t-reaction time (min).

C_t-concentration of pollutant at time t (mg/L).

C_0-equilibrium concentration of pollutant after dark adsorption-desorption equilibrium (mg/L).

A_t - the absorbance of the contaminant at time t.

A_0-absorbance of the pollutant after dark adsorption-desorption equilibrium.

4.2 Photocatalytic properties

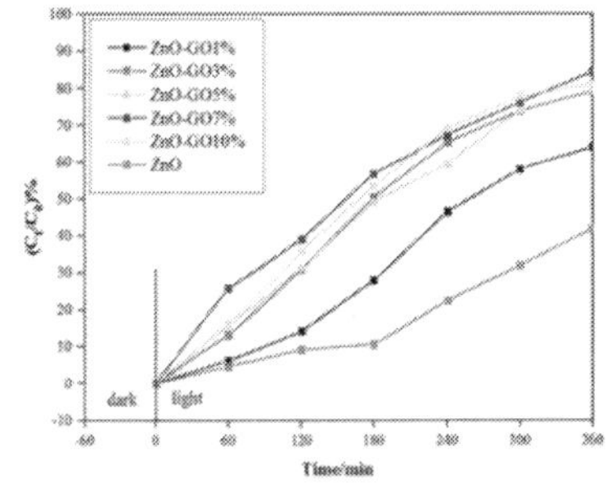

Fig. 4. Effect of different GO concentrations on the photocatalytic performance of the catalyst.

The photocatalytic activity of rhodamine B was tested by desorption of the sample under UV light. The reaction was preceded by adsorption in a dark box for 30 min and after reaching adsorption equilibrium, the UV light source was switched on. Figure 4 shows the photocatalytic activity of the different samples[7]. It can be seen that the ZnO nanopowder has a certain degradation effect on rhodamine B, with a degradation rate of 41.96% at 360min of light exposure. From the photodegradation curves of the ZnO-GO composites for rhodamine B, more GO content does not necessarily mean that the photocatalytic performance of the composites is good, probably because too much GO content in the composites will instead affect the absorption of UV light by ZnO, thus affecting the photoexcitation and catalytic activity of the ZnO particles. The proper introduction of GO can make the composite adsorb organic dyes well and can promote the effective separation of ZnO photogenerated carriers and reduce the composite rate. This shows that the ratio of ZnO and GO in the composite has a very strong influence on photocatalysis. Among them, ZnO-GO 7% has the best photocatalytic performance, with a degradation rate of 84.42% at 360 min of light.

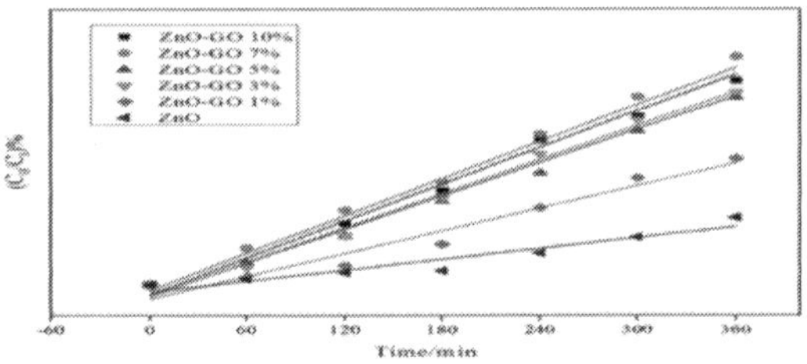

Fig. 5. ln(C0/Ct)-t linear fit curves for catalyst samples with different GO concentrations.

Table 1. Linear correlation coefficients between reaction rate constants and regression equations for different catalysts.

Sample name	Regression equation	K	R^2
ZnO	y=0.0995x-0.1077	0.00147	0.928
ZnO-GO1%	y=0.2975x-0.1769	0.00307	0.9951
ZnO-GO3%	y=0.2906x-0.1733	0.00464	0.9887
ZnO-GO5%	y=0.2069x-0.2134	0.00500	0.9791
ZnO-GO7%	y=0.3114x-0.0813	0.00504	0.9909
ZnO-GO10%	y=0.316x-0.1506	0.00453	0.9895

Figure 5 shows the linear fitting curve of $\ln(C_0/C_t)$-t for ZnO-GO 7%. It can be seen that $\ln(C_0/C_t)$-t shows a good linear correlation, indicating that the photocatalytic degradation of rhodamine B is an apparent primary reaction. The rate constants and linear correlation coefficients of the regression equations are shown in Table 1, which shows that ZnO-GO 7% had the best catalytic effect with the highest rate constant of 0.00504. The kinetic fitting results are consistent with the degradation efficiency results in Figure 4. The linear correlation coefficients of the regression equations were all greater than or close to 0.99, indicating a good regression effect.

5 Conclusion

Zno nanoparticles were obtained by simple hydrothermal method, and go and GO were obtained by modified Hummer method, and ZnO-GO composites with different GO concentrations were prepared by mixing ZnO nanoparticles with GO. Here we study the efficiency of rhodamine B as a pollutant for photocatalytic degradation by using a xenon lamp greater than 400 nm as the unique light source.The experimental results showed that ZnO-GO 7% had the best catalytic effect with 84.42% degradation rate when the light time was 360min. The composite material is simple to prepare and has certain practical application value.

References

1. Ling Sun. Structure and synthesis of graphene oxide[J]. Chinese Journal of Chemical Engineering. 2019(10)
2. ANDrew T.Smith,Anna Marie La Chance,Songshan Zeng, Bin Liu,Luyi Sun. Synthesis, properties, and applications of graphene oxide/reduced graphene oxide and their nanocomposites[J]. Nano Materials Science. 2019(01)
3. Xiao-Jun Shen, Shu Yang,Jian-Xiang Shen,Jun-Li Ma,Yin-Qiu Wu,Xiao-Ling Zeng,Shao-Yun Fu.Improved mechanical and antibacterial properties of silver-graphene oxide hybrid/polylactid acid composites by in-situ polymerization[J]. Industrial Crops & Products . 2019
4. Cheng Zhang,Zhigao Wang,Yang Li,Yijie Yang,Xingrong Ju,Rong He. The preparation and physiochemical characterization of rapeseed protein hydrolysate-chitosan composite films[J]. Food Chemistry. 2019
5. Suhaili Shamsi,Nurhuda Elias,Seri Narti Edayu Sarchio,Faizah Md Yasin. Gallic Acid Loaded Graphene Oxide Based Nanoformulation (GAGO) as Potential Anti-bacterial Agent against Staphylococcus aureus[J]. Materials Today: Proceedings. 2018
6. Sunil Meti,Mohammad R. Rahman,Md. Imteyaz Ahmad,K. Udaya Bhat. Chemical free synthesis of graphene oxide in the preparation of reduced graphene oxide-zinc oxide nanocomposite with improved photocatalytic properties[J]. Applied Surface Science . 2018
7. Zhang Jiaqi,Li Jin,Liu Xiangyu. Ternary nanocomposite ZnO-g-C3N4-Go for enhanced photocatalytic degradation of RhB[J]. Optical Materials,2021,119:

The potential of using corn straw biochar to compensate the loss of available soil organic matter

Leisan Akhmetzyanova[1,*], *Ivan* Kuritsin[2], *Marat* Zinnatov[1], *Alexander* Gordeev[1], *Alina* Kamalova[1], and *Kamalya* Karamova[1]

[1] Institute of Environmental Sciences, Kazan Federal University, 420018 Kazan, Russia
[2] LLC "Reculto", 420097 Kazan, Russia

Abstract. The process of migration of organic carbon from the biochar into the substrate and the aqueous phase was simulated. The process of organic carbon migration in passive and washing mode was evaluated. Part of the organic carbon was washed with water and passed into the aqueous phase, the other part remained in the mass of biochar. Biochar can be a source of available organic carbon, providing nutrition to rhizosphere microorganisms that improve the mineral nutrition of plants, both in the short and long term.

1 Introduction

The unreturnable removal of organic carbon from the soil with agricultural products makes the problem of soil depletion worse every year. The search and development of measures for environmentally safe farming is a key issue of modern agriculture.

Biochar has many advantages, since it improves the mechanical structure of the soil. Biochar produced from agricultural waste is also a clean and safe source of carbon. In recent years, there is a number of positive results from the introduction of biochar into the practice of agriculture in different countries [1-4]. The production of fertilizers from biochar contributes to solving the problem of storage and processing of waste from agriculture, animal husbandry, food industry and others [5]. Biochar, made on the basis of biomass, is used for sequestering carbon, sorbent for heavy metals, etc. [6, 7].

Currently, there is not enough research on the use of biochar in Russia, there is no practice of its use in agriculture. Most often, biochar is considered only as an additive that improves the physical and mechanical properties of soils, while its effect on the agrochemical properties of soils is not analyzed. There are also no exhaustive data about the biochar carbon migration process along the soil profile, and the mechanisms of interaction of biochar components with the soil. The mechanism of migration of the biochar along the soil profile needs additional research.

The aim of this work was to quantify the migration ability of the water-soluble carbon fraction of a biochar during laboratory modeling of the leaching process.

[*] Corresponding author: leisan-ksu@mail.ru

2 Materials and methods

The object of the study is a biochar of corn straw obtained at a temperature of 400 degrees Celcius. At the first stage of the experiment, laboratory modeling of the process of passive migration of water-soluble organic carbon (WSOC) from biochar into water was performed. Dry biochar (15 g) was placed in conical flasks and 150 ml of distilled water was added, stirred on a rotator for 8 hours. Then the flasks were kept in a quiet state at temperature between 21-25 degrees Celcius without exposure to direct sunlight. The exposure of the biochar in water was 3, 7, 14, 21 and 28 days. After the specified exposure periods, the determination of WSOC in an aqueous extract was carried out.

At the second stage, laboratory modeling of the process of vertical migration of WSOC of the biochar was carried out. For this 3 stainless steel box columns with a side of the walls of 30 mm a height of 200 mm were used. Aluminum oxide (Al_2O_3) was used as a substrate in the experiment. Biochar (50 g) and aluminum oxide were poured into the column in a ratio of 1:3 (Fig. 1) [8, 9].

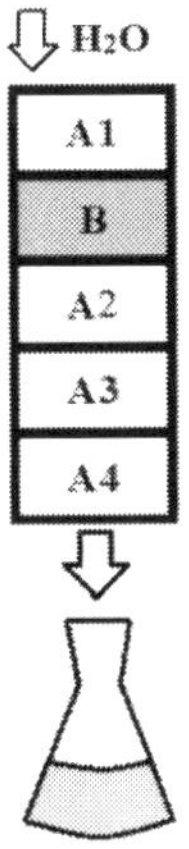

Fig. 1. The design of vertical migration columns.

The columns were irrigated from above with distilled water in 150 ml portions. The washings were performed three times, waiting for the eluate to be completely excavated. Thus, three consecutive portions of the extract were obtained. Portions of eluate were collected in glass flasks. Aqueous extracts were examined for the content of WSOC by bichromate oxidation in a sulfuric acid medium with a spectrometric termination.

After the end of the washing, the distribution of the water-soluble carbon fraction in the migration profile was evaluated. Samples of biochar (B) and aluminum oxide (A1, A2, A3, A4) were dried to a constant mass. In the samples taken from different layers of the column after washing, the content of total, organic and inorganic carbon was determined. Determination of the total carbon content was carried out according to the Dumas method using the vario MAX cube elemental analyzer (Elementar, Germany). The content of individual carbon phases (organic and inorganic) was determined using an elemental analyzer with thermogradient analysis Leco r-612 (Leco Instruments, USA).

All parameters were measured not less than 3 times. Microsoft Excel software was used to carry out the statistical processing. The reliability of the mean differences was assessed using the Student's coefficient ($P<0.05$).

3 Results and discussions

On the 3 day of first stage the experiment, the concentration of WSOC in the eluate was 8.6 mg/ml (fig.2). On the 7 day a significant increase was observed. The concentration of

WSOC in the extract was 49 mg/ml. A week later, an increase in concentration to a value of 61 mg/ml was also noted. Subsequent measurements did not show a significant change in the value of the extracted carbon during the experiment. Thus, the greatest efficiency of leaching of WSOC was observed on the 14th day of the experiment. The exposure time insignificantly changed the quantity of organic carbon extracted by water.

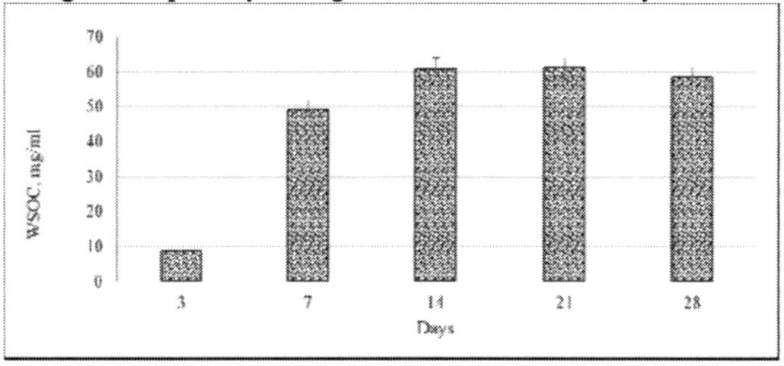

Fig. 2. The concentration of water-soluble carbon in the eluate.

The results of measurements of the concentration of organic carbon in the eluate at the second stage of the experiment are shown in Figure 3. The concentration of WSOC in the first portion of the eluate had values in the range of 7-9 mg/ml. The highest concentration was observed in the second portion of the eluate and had an average value of 17.2 mg/ml. The portion of eluate collected after the third washing had an average value of 7.8 mg /ml.

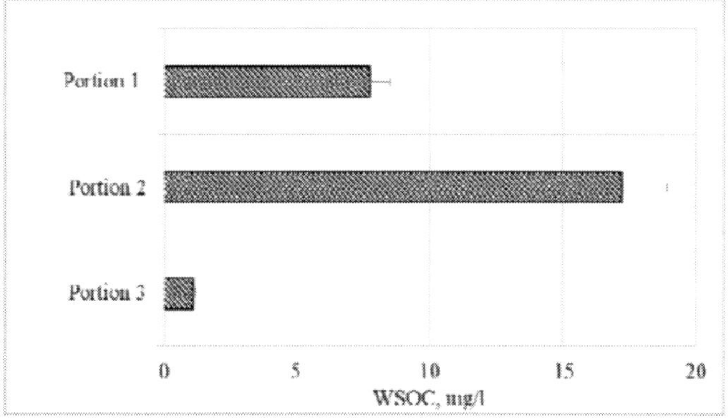

Fig. 3. The concentration of organic carbon in the eluate collected after washing the columns.

The first portions of water moisten the contents of the columns. The removal of WSOC from the mass of the biochar by subsequent portions of water becomes easier. This explains the increase in its concentration in the second portion of the eluate. Subsequent portions of water wash out the remaining amount of water-soluble compounds and in the third portion the concentration is significantly reduced. Accordingly, it can be assumed that the introduction of biochar into the substrate, as a source of organic carbon, will have greater efficiency at the initial stage, losing it during subsequent rinsing with water.

After washing the columns the carbon content in their various parts was disassembled and analyzed. The total carbon content in the biochar samples before the experiment (the initial biochar) was about 61%, 59% of which - organic carbon, about 2% - inorganic. Results shown in the Table 1.

Of greater interest, from the point of view of fertilizing properties, is the content of water-soluble organic carbon in different layers of the substrate profile and in the aqueous phase after washing [10]. Dissolved organic carbon is extremely important for the functioning of terrestrial ecosystems — especially for the formation of the chemical composition of soils and their genesis. [11], migration of nutrients and metals, as well as a substrate for soil microorganisms [12]. Due to the high content of available carbon, the nutrition of rhizosphere microorganisms that improve the mineral nutrition of plants is provided.

Table 1. Carbon content in various parts of the columns after washing.

Element	Total carbon, $\%_w$	Organic carbon, $\%_w$	Inorganic carbon, $\%_w$
Initial biochar	61,18±1,91	59,185±1,85	1,995±0,06
A1	0,040±0,00166	0,039±0,0016	0,00158±0,00006
B	37,603±1,17	36,583±1,14	1,02±0,032
A2	0,057±0,00166	0,055±0,0016	0,00224±0,00006
A3	0,054±0,00166	0,052±0,0016	0,00194±0,00006
A4	0,062±0,00166	0,060±0,0016	0,00193±0,00006

The residual amount of organic carbon in the biochar layer (B) after washing was 36.6%. Thus, part of the organic carbon was washed out with water through the substrate (fig.4). The total carbon content in the aluminum oxide samples in the layer above the biochar (A1) was 0.04%, the largest share in this case was organic carbon (0.039%). The content of organic carbon in the layers of aluminum oxide under the biochar is 1.4 times higher than in the upper one. Consequently, this component of biochar migrates down the profile, and lingers in the substrate, albeit in small quantities.

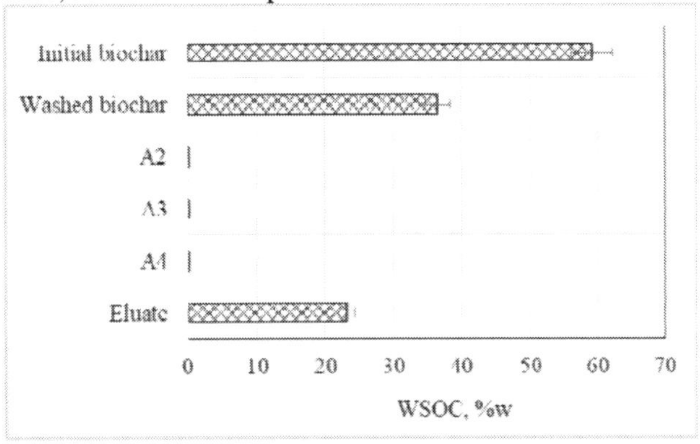

Fig. 4. The concentration of organic carbon in different layers of columns.

A small amount of carbon has been fixed on the substrate, under natural conditions, an increase in this value should be expected [13]. Organic matter from the biochar in the soil will be retained not only mechanically, but also at the level of chemical interaction with soil organic matter determined by its content and structure. The method of agricultural tillage, soil structure and filtration coefficient also are important.

The loss of organic carbon from the mass of the biochar after washing are about 23%. Consequently, corn straw biochar consists of almost a quarter of WSOC and can act as a source of available organic carbon. Which, under conditions of sufficient moisture, can become an additional source of nutrients for soil microorganisms. A. Obia found out that

19% of carbon from biochar migrated to the lower layers of the soil. The penetration depth is 30-40 cm from the biochar layer and increases over time [12].

It is shown that the components of the biochar are capable of vertical migration along the soil profile (30-50 cm). The authors attribute this to direct leaching and infiltration of biochar in soluble or solid forms due to a large amount of precipitation, or ingestion and bio-displacement by earthworms [11].

Analyzing the data obtained, it can be assumed that the corn straw biochar consists of 59% organic carbon. On the one hand, it can serve as a fast source of organic matter (23 %WSOC). On the other hand, part of the organic carbon remains in the mass of the biochar and has the potential, due to the activity of living organisms, to slowly pass into the substrate and further into the soil solution, thereby enriching it with available organic matter [14, 15]. Thus, the use of corn straw biochar can have a beneficial effect on the nutrition conditions of rhizosphere microorganisms and, as a consequence, plants by improving the properties of the soil substrate.

3 Conclusion

The main part of the carbon of the biochar from corn straw belongs to the total organic carbon, the proportion of inorganic carbon in the samples is insignificant.

Passive exposure of the biochar in water showed the greatest efficiency of leaching of WSOC from biochar was observed on the 14th day of the experiment. The exposure time insignificantly changed the quantity of organic carbon extracted by water.

The water-soluble organic carbon of the corn straw biochar is able to migrate quite actively along the vertical profile into the aqueous extract. Carbon capture on an inert substrate is insignificantly.

Biochar produced from agricultural waste can be a clean, safe source of available organic carbon and can have a beneficial effect on the soil biota.

References

1. J. Lehmann, S. Biol. Bioch., **43**, 1812-1836 (2011)
2. S. Jeffery, Agr. Ecosyst. Env., **144**,175-187 (2011)
3. D.N. Mulcahy, J. Arid. Env., **88**, 222-225 (2013)
4. P. Blackwell, G. P. Riethmuller, Earthscan, (2009)
5. S.P. Galinato, J.K. Yoder, D. Granatstein, Enrg. Pol., **10**, 6344–6350 (2011)
6. K. Weber, P. Quicker, Fuel. Elsevier., **217**, 240–261 (2018)
7. P. Galitskaya, L. Akhmetzyanova, S. Selivanovskaya, Biogeosci., **13**, 5739-5752 (2016)
8. A. Giniyatullin, Res. J. Pharm. Biol. Chem. Sci., **4**, 194-201 (2015)
9. McKnight D.M., Env. Sci. Tech., **7**, 1388-1396 (1992)
10. F. Santos, S. Biol. Bioch., **51**, 115-124 (2012)
11. Y. Fang, Plos One, **43**, 48 (2015)
12. A. Obia, S. Till. Res., **40,** 30-34 (2017)
13. A. Keith, Env. Sci. Tech. **45**, 9611-9618 (2011)
14. Y. Kuzyakov, S. Biol. Bioch., **32,** 1485-1491 (2000)
15. I. J. Umaru, B. Samling, U.H. Aduwamai, As. J. Bioch. Gen. Mol. Biol., **10**, 12-15 (2019)

Characteristics of wind turbine wake and particle transport in aeolian sand environment

Mandi Sha[1*], *Xianbin* Xiao[2], *Wu* Qin[3], and *Zongming* Zheng[4]

College of new energy, North China Electric Power University, 102206 Beijing, China

Abstract. For the wind turbine wake field in the aeolian sand environment, the multiphase flow model is used to simulate the wind sand inflow and the wake of the wind turbine is simulated and calculated in combination with the actuation line model based on the OpenFOAM-v2006. The effects of clean air inflow and wind sand inflow on the wind turbine wake field are compared, including speed loss and wake structure. Equally the particle transport in the wake of wind turbine is analyzed. The results show that the velocity of wind turbine wake recovers faster in the aeolian sand environment, the overall spatial structure is sinking, and the particles will settle in advance and distribute in strips.
Keywords: ALM, OpenFOAM, MPPICFoam, LES.

1 Introduction

In Northwest and North China, the wind resource rich area highly coincides with the aeolian sand area, and the influence of wake effect must be considered in the micro site selection and layout of wind turbines [1]. Deshun Li analyzed the influence of wind sand inflow on blade wear through Fluent [2-3]. Congxin Yang analyzed the wake characteristics of wind turbine under clean inflow by actuation line model simulation [4]. Due to the solid modeling method is limited by the amount of grid calculation, and cannot simulate multiple wind turbines. As a relatively accurate wake model, the actuator line model does not need solid modeling and has a small amount of calculation. At present, its application is mainly concentrated in single-phase flow system, and there is little research on the application of actuation line in multiphase flow system.

2 Theory and model

2.1 Multiphase flow model(MP-PIC)

As a particle tracking model, MP-PIC model describes the gas phase by Euler method, calculates the relative motion with the discrete phase by Lagrange method, and processes the collision between particles by calculating the pressure gradient force to reduce the amount of calculation.

*Corresponding author: smd18801096587@163.com

Gas phase continuity equation:

$$\frac{\partial}{\partial t}(\varepsilon_g\rho_g) + \nabla(\varepsilon_g\rho_g\mathbf{u_g}) = 0 \tag{1}$$

Gas phase momentum equation:

$$\begin{cases} \frac{\partial}{\partial t}(\varepsilon_g\rho_g\mathbf{u_g}) + \nabla\cdot(\varepsilon_g\rho_g\mathbf{u_g}\mathbf{u_g}) = -\varepsilon_g\nabla p + \nabla\cdot\boldsymbol{\tau}_\mathbf{g} + \varepsilon_g\rho_g\mathbf{g} - \sum_{p=1}^{n_T} n_p\frac{V_p}{V_g}\beta_p(\mathbf{u_g}(\mathbf{x_p}) - \mathbf{u_p}) \\ \boldsymbol{\tau}_\mathbf{g} = \mu_g\left[\nabla\mathbf{u_g} + (\nabla\mathbf{u_g})^T\right] - \frac{2}{3}\mu_g\nabla\mathbf{u_g}\mathbf{I} \end{cases} \tag{2}$$

Where: ε_g is the porosity, ρ_g is the gas phase density, $\mathbf{u_g}$ is the gas phase velocity, $\mathbf{u_p}$ is the particle velocity, $\mathbf{x_p}$ is the particle displacement. n_T is the total number of calculated parcels in the cell, n_p is the number of real particles in the calculated parcels, $\mathbf{I}$ is the unit tensor, $\mathbf{g}$ is the gravitational acceleration, β_p is the drag coefficient, $\boldsymbol{\tau}_\mathbf{g}$ is the corresponding air force, V_p is the volume of particles in the cell, V_g is the fluid volume in the cell.

Equation of motion of solid phase:

$$\begin{cases} \frac{d\mathbf{x_p}}{dt} = \mathbf{u_p} \\ \frac{d\mathbf{u_p}}{dt} = -\frac{\nabla p}{\rho_s} - \frac{\nabla p_s}{\varepsilon_s\rho_s} + \mathbf{g} + \frac{\beta_p}{\rho_s}[\mathbf{u_g}(\mathbf{x_p}) - \mathbf{u_p}] \\ p_s = \frac{p_s^*\varepsilon_s^\alpha}{\varepsilon_{s.max} - \varepsilon_s} \end{cases} \tag{3}$$

Where: ρ_s is the solid density, $\varepsilon_{s,max}$ is the maximum solid holdup, p_s^*, α is the inherent parameter of the model, generally [5] $10 < p_s^* < 100, 2 < \alpha < 5$.

Ergun-WenYu drag model is selected as the drag model for calculation. The drag model combines Wen Yu model and Ergun model [6]. Among them, the calculation of the drag force on particles in the flow field is as follows:

$$\mathbf{F_p} = \frac{3}{4}m\alpha_p\rho_c\frac{C_D}{d_p}(\mathbf{u_g} - \mathbf{u_p})|\mathbf{u_g} - \mathbf{u_p}| \tag{4}$$

2.2 Wind turbine actuator line model (ALM)

The actuator line model divides the blade into multiple micro elements according to BEM(blade element momentum theory). When the fluid flows through the blade, the aerodynamic force on each section is obtained by interpolation calculation according to the two-dimensional airfoil aerodynamic characteristic parameters of each blade element section, and the calculated volume force is substituted into the flow field to calculate the information of the surrounding flow field. The aerodynamic volume force is calculated as follows:

$$\mathbf{f_{ae}} = (\mathbf{L}, \mathbf{D})^T = \frac{1}{2}\rho\mathbf{W}^2c(C_L\mathbf{e_L}, C_D\mathbf{e_D})^T$$

Where, $\mathbf{W}$ is the local wind speed, in $m\cdot s^{-1}$; c is the chord length, in m; ρ is the local air density, in $kg\cdot m^{-3}$; C_L is the airfoil lift coefficient, C_D is airfoil drag coefficient; $\mathbf{e_L}$, $\mathbf{e_D}$ are unit vectors in the direction of lift and drag respectively.

2.3 Large eddy simulation method (LES)

In order to capture the vortex system structure in the wake, the turbulence model adopts large eddy simulation (LES). In the LES turbulence model, the Smagorinsky model overestimates the turbulent viscosity, so the solution is relatively stable and the calculation speed is fast. It is the most widely used LES model at present.

2.4 Physical model and inflow conditions

The simulation target in this paper is the wind turbine experiment of aeolian sand environment based on the Environmental Wind Tunnel of the Key Laboratory of Western Disaster and Environment of Lanzhou University. Due to the large calculation demand of the particle tracking model, the numerical simulation model is the model unit of wind tunnel test, and the calculation object is the reduced scale model. Based on this, the gas-solid flow characteristics of the wind turbine wake of the scale model are simulated and analyzed.

The calculation area ranges from 2D in front of the wind turbine to 20D behind the wind turbine, with a size of $1.3\text{m} \times 1.4\text{m} \times 10.8\text{m}(W \times H \times L)$. The grid densification area ranges from 0.5D in front of the wind turbine to 15D behind the wind turbine, with a size of $0.6\text{m} \times 0.6\text{m} \times 6.82\text{m}(W \times H \times L)$ and adopt SnappyHexMesh for grid encryption. The total number of grids is 5 million. **Fig. 1** shows the grid division used in the calculation, where A is the layout position of the wind turbine, B is the grid densification area, and C is the fluid domain of the calculation.

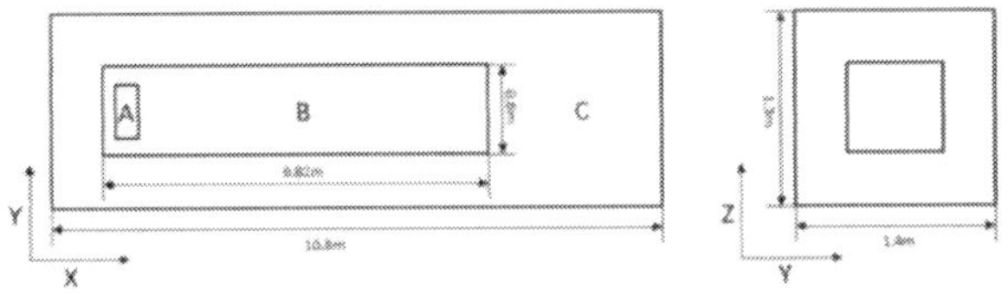

Fig. 1. Diagrammatic sketch of meshing.

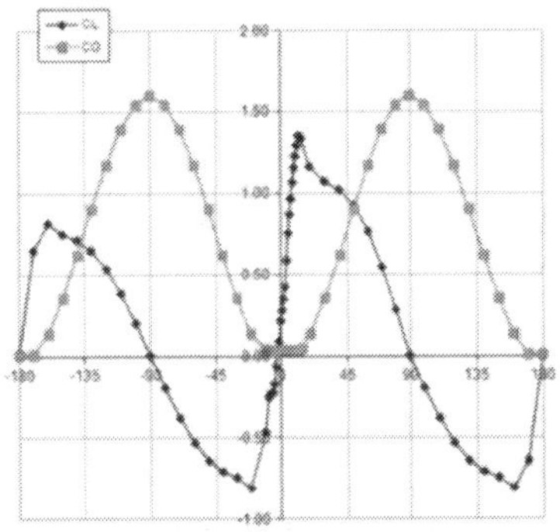

Fig. 2. SG5060 Airfoil lift and drag coefficient.

The boundary settings and wind turbine parameter settings are shown in **Table 1.**

Table 1. Basic parameters.

Parameter	Numerical value	Unit
Incoming flow velocity	8	m/s
Tip speed ratio	4	/
Number of blades	3	/
Airfoil	SG6050	/
Diameter of wind turbine	0.44	/
Hub height	0.675	m
Particle size	70	um
Number of parcels per second	17400321	/
Number of particles per parcel	1000	/
Particle density	2560	kg/m^3

According to the blade element momentum theory, the aerodynamic design of wind turbine blade is carried out. Different Airfoils are selected for different spanwise lengths, and the lift coefficient and drag coefficient are interpolated within the angle of attack range of 0 ° - 180 °. The results are shown in the **Fig. 2**.

3 Results and discussion

This paper simulates and analyzes the gas-solid flow characteristics in the test section, compares the wind turbine wake flow field of clean air and the wind turbine wake flow field under the incoming sand, and studies the influence of particles on the temporal and spatial evolution of wake and the influence of wind turbine wake on particle transport.

3.1 Structure and velocity recovery of wind turbine wake

Under the inflow condition of 8m/s under the rated working condition of the wind turbine, the wake velocity distribution curve of the wind turbine presents an M-shaped distribution and recovers with the distance. Compare the velocity distribution of wind turbine wake field under clean air and sand, as shown in Figure 3 and select the position 1D, 5D, 10D and 15D behind the wind turbine to make the velocity distribution of XY section and XZ section, as shown in **Fig. 4**.

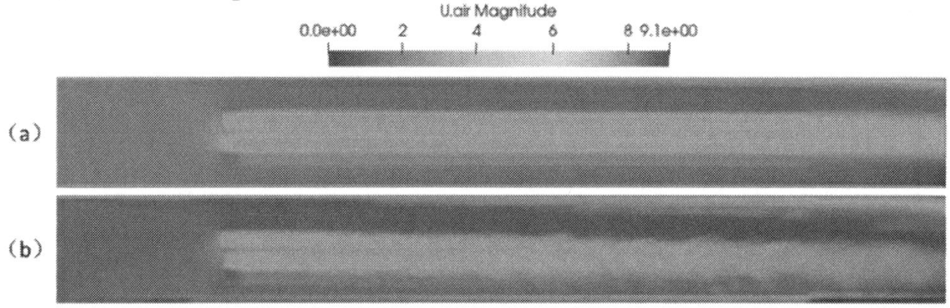

Fig. 3. Velocity cloud diagram of wind turbine wake field (a) clean air inflow (b) wind sand inflow.

For the wind sand flow, the settlement effect caused by the gravity of the particles will be transmitted to the wind turbine wake through the drag force between the gas phase and the particle phase, resulting in the downward trend of the wind turbine wake in space. **Fig. 4** (b) shows the flow field velocity cloud diagram of XZ section.

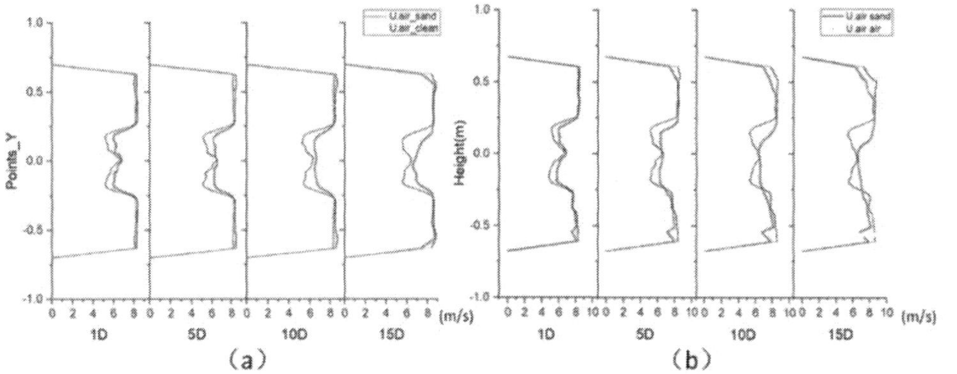

(a) XY section (b) XZ section

Fig. 4. Comparison of wake velocity under the condition of clean air inflow and wind sand inflow.

The minimum value of the velocity profile of each section in **Fig. 4** are selected as the characteristic quantity for comparative analysis, and the results are shown in **Table 2**. Due to the influence of particle movement, the air flow is more disordered and the wake of the

 https://doi.org/10.1051/e3sconf/202234101003

wind turbine will recover relatively quickly.

Table 2. Comparison of wind turbine wake loss under different incoming flows.

Parameter	1D	5D	10D	15D
Clean air	5.11258	5.15867	5.34404	5.54289
Velocity loss	0.36092	0.35516	0.33199	0.30713
Wind sand flow	5.89507	6.11965	6.39141	6.51209
Velocity loss	0.26311	0.23504	0.20107	0.18598

3.2 Influence of wind turbine wake on particle transport

When the air flows through the plane of the wind turbine rotor, the wind machine absorbs energy from the flowing air, resulting in the loss of air energy and the decrease of speed. At this time, due to the drag force of the gas-phase fluid, the particles will also decelerate accordingly. Due to the constant existence of gravity and the weakening of the upward velocity of the square of water, the particles will settle in advance. Due to the wind turbine wake structure is three periodically changing spirals, the particles will show a banded distribution during deposition due to the influence of the wake.

The number of particles from the surface to 0.44m above the ground is analyzed. **Fig. 5** shows that there will be a peak value of the number of particles every 1m along the incoming flow direction, which is similar to the periodic change of the wind turbine wake in space. Compared with the partial velocity of the flow field in the Y direction at this position, shown in **Fig. 6,** the changing value is similar to the distribution of the number of particles, It shows that the wind turbine wake can promote the strip deposition of particles.

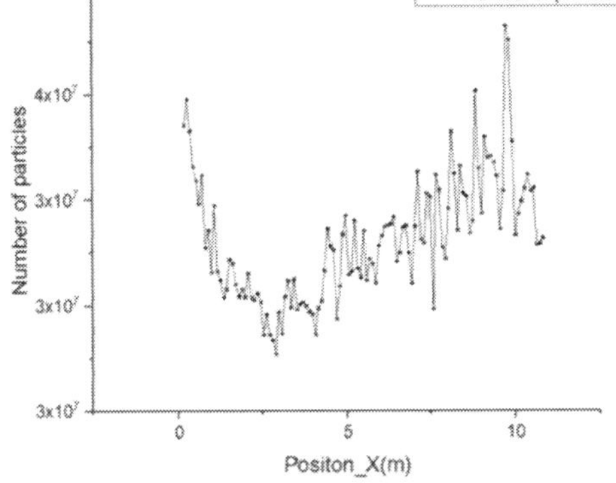

Fig. 5. Number of particles from ground to 0.44m above.

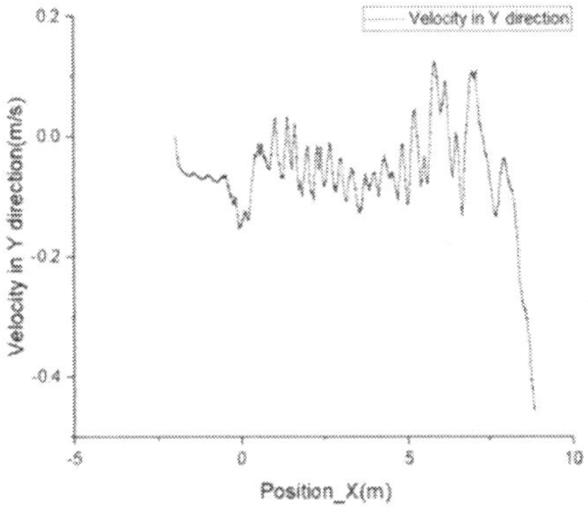

Fig. 6. Velocity of Y direction.

4 Conclusion

In this paper, multiphase flow model and actuation line model are coupled based on OpenFOAM-v2006 in order to simulate the temporal and spatial changes of wind turbine wake under clean air and wind sand inflow. The loss and morphological structure of wind turbine wake are analyzed, and the influence of wind turbine wake on particle transport is concerned.

1) Because of the movement of particles in the flow field, the wind turbine wake is more disturbed, the turbulent kinetic energy increases, and the wake recovery increases.
2) Due to its own gravity, particles are transmitted to the air through the drag force between the fluid and solid phases, resulting in the sinking of the wind turbine wake field in the spatial structure.

3) Under the periodic rotation of wind turbine blades, the change of wake has periodic characteristics. Therefore, the particles will slow down due to the influence of wind turbine wake during the settlement process, resulting in early deposition and strip distribution.

As a result of the simple grid required by the actuation line model, the wake simulation of multiple units can be expanded in the future to study the temporal and spatial evolution of the wake structure of the wind farm in the aeolian sand environment, so as to optimize the layout and improve the power generation efficiency.

References

1. Energy - Wind Energy; Studies from Technical University Further Understanding of Wind Energy (Analytical model for the power-yaw sensitivity of wind turbines operating in full wake)[J]. Energy Weekly News (2020)
2. Li Deshun, Chen Xia, Li Yinran, Guo xingduo, Wang Yae Study on the evolution process of wind sand erosion wear of wind wing [J] Gansu Journal of science **32**, 05 (2020),32(05):63-68+87.DOI:10.16468/j.cnki.issn1004-0366.2020.05.013.
3. Li rennian, Zhao Zhenxi, Li Deshun, Li Yinran, Chen Xia, Yu Jiaxin Effect of wind sand on the flow around wind wing and its aerodynamic performance [J] Journal of agricultural engineering 34, 14(2018),34(14):205-211+303.
4. Yang Congxin, Zhang Yaguang, Zhang xuyao, He pan Study on wake characteristics of wind turbine based on large eddy simulation [J] Mechanical design and manufacturing (2021)
5. Auzerais F M, Jackson R, Russel W B. The resolution of shocks and the effects of compressible sediments in transient settling [J]. J. Fluid Mech (2006)
6. GIDASPOW D. Multiphase flow and fluidization: continuum and kinetic theory descriptions [M]. Academic press **42** (1994)

Investigation of cleaning effect for airport runway lamps by using baking soda powder

Yang Zhao[1,2], *Xiaohong* Ge[1,2,*], *Xing* Wang[1,2], *Ke'an* Yue[1,2], *Yi* Chen[1,2], and *Jinhuo* Wang[1,2]

[1] Key Laboratory of Functional Materials and Application of Fujian Province, Xiamen University of Technology, Xiamen 361024, China;
[2] Xiamen Engineering and Technology Research Center of High Efficiency and Precision Intelligent Manufacturing, School of Materials Science and Engineering, Xiamen University of Technology, Xiamen 361024, China;

Abstract. In order to ensure the bright effect of airport navigation lights, dry powder cleaning technology was adopted in this paper to improve the efficiency of removing sticky dirt on the lamp surface. The atomization scheme by mixing dry powder with water was designed through deliberately selecting the appropriate injection nozzle structure to reduce impact of powder pollution on the environment. Both simulation analysis and experimental testing showed that the overall cleaning effect and the atomization scheme were feasible, which reduce the dust pollution by more than 90%.

1 Introduction

In recent years, with rapid development of the aviation industry, the time for cleaning and maintenance of navigation lights at the airport (as shown in Fig.1) has become shorter and shorter. In this paper, the dry powder cleaning method is adopted for research and application.

Dry powder abrasive consists of sodium bicarbonate powder, commonly known as baking soda, which is an alkaline abrasive with water solubility of 7.8g (18℃) in water.

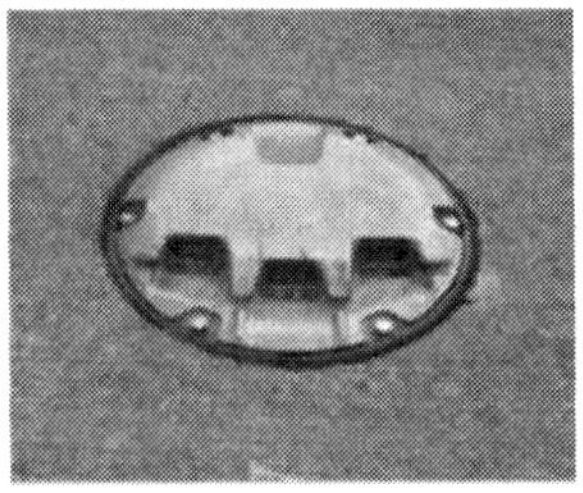

Fig.1. An embedded airport navigation lamp.

Hu Weiwei [1] from Xiamen University of Technology used solid carbonate powder as the cleaning medium, commonly known as a dry powder, and optimized the cleaning parameters. Wang et al. [2] studied the cleaning effect of dry powder cleaning technology on airport

*Corresponding author: xhge@xmut.edu.cn

navigation lights. Gao et al. [3], Cheng [4] also used dry powder cleaning technology successively to study the effect of dry powder cleaning navigation lights.

However, dust and other issues also exist in the process of cleaning, this paper is thus to investigate the cleaning equipment by optimizing water atomized parameters, aiming at improving the airport navigation lights with high cleaning efficiency. At the same time, it is to reduce the harm of dust to the human body in dry powder cleaning.

2 Mechanism of atomization dust suppression

The theory of liquid jet atomization in transverse airflow was first proposed by Chen et al[5]. The atomization process mainly includes two aspects: primary crushing process and secondary crushing process. By fluid instability analysis, it can be seen that the liquid jet leaves the nozzle as a complete liquid column. Because of the axial instability, the developed liquid column eventually decomposes into droplets and liquid blocks, which is called the column splitting process. (Fig.2)

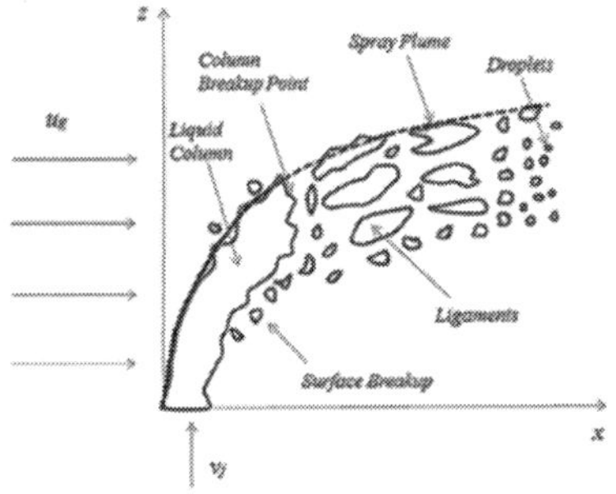

Fig.2. Atomization process of a liquid jet in transverse flow.

The most commonly used liquid jet-breaking standard for transverse airflow is proposed by Ohnesorge[6].

Based on previous studies, Reitz[7] attempted to resolve some of the uncertainties on the Ohnesorge chart. His analysis is based on an interpretation of data from diesel spray. According to the Reitz mechanism, as shown in Fig.3, as the liquid injection rate gradually increases, the following four decomposition modes are obtained: Rayleigh and broken, the first disruption, the second wind induced, and the atomization.

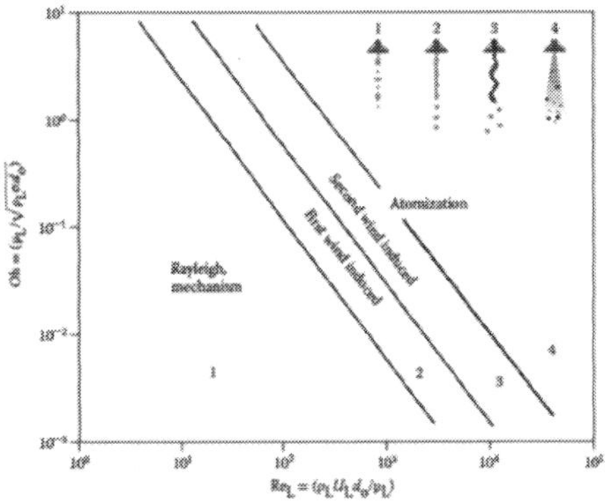

Fig. 3. Reitz mechanism broken.

3 Simulation analysis of atomization

3.1 atomization modeling and parameter setting

To simulate the atomization and breakage process and jet characteristics of a liquid jet in transverse airflow, CLSVOF method has been used to study the breakage process and breakage position characteristics of a liquid jet in the transverse airflow. CLSVOF method is combined with Level Set and VOF method. The analysis software Fluent was used to determine the calculation domain and boundary settings inside the nozzle, as shown in Fig.4. The wall surface of the injection flow was set as the non-slip boundary, and the other boundaries were set as the slip boundary. The computational domain grid was divided into unstructured grids with a uniform density distribution. To prevent pseudo-breakage of the liquid phase caused by the inhomogeneity of the computational domain grid, domain size values were calculated as shown in Table 1.

Table 1. Calculate the dimensions of the domains.

Area	Size(mm)
Inlet diameter of the air	9 mm
Length of the calculation domain	0mm ≤ z ≤ 60mm
nozzle diameters	1 mm

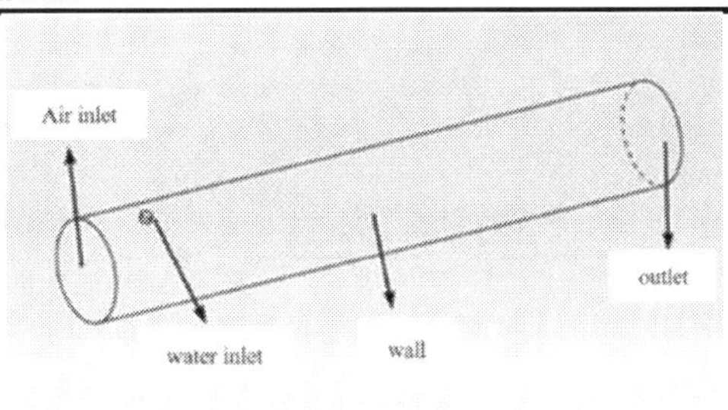

Fig.4. Crushing process simulation study calculation area

Because of the high velocity of transverse airflow, it can be regarded as compressible fluid, so it was calculated based on the pressure solver. The turbulence model adopted the LES model, which can accurately capture the air-liquid interface and check the crushing process. The boundary conditions are shown in Table 2, and the physical parameters remain unchanged in the calculation process.

Table 2. Boundary condition setting of simulation.

Initial condition	Parameter
Air inlet	Pressure inlet: 0.2 MPa -0.5MPa
Irrigation-water inlet	Pressure inlet: 0.5 MPa
Outlet	Pressure outlet: 1.01325 MPa

3.2 Jetting characteristics of a liquid jet in the nozzle

Injecting depth is an important parameter to evaluate the mixing characteristics of the liquid jet and free air, which can affect the atomization efficiency. The injection depth has a great influence on the liquid jet atomization and the pressure change in the pipe.

It can be seen from Fig. 5 that the crushing position is affected by the air inlet pressure. As the air inlet pressure increases, the crushing position moves forward. When the air inlet pressure is 0.5MPa, the crushing occurs first. Under the condition of constant water injection pressure, it can be seen from Fig. 6 that the injection depth decreases gradually with increasing air inlet pressure. When the air inlet pressure is 0.5MPa, the injection depth is the minimum. Under the influence of high-pressure airflow, the liquid jet rapidly decomposes into small droplets of different sizes, which are mainly concentrated in the front area of the jet.

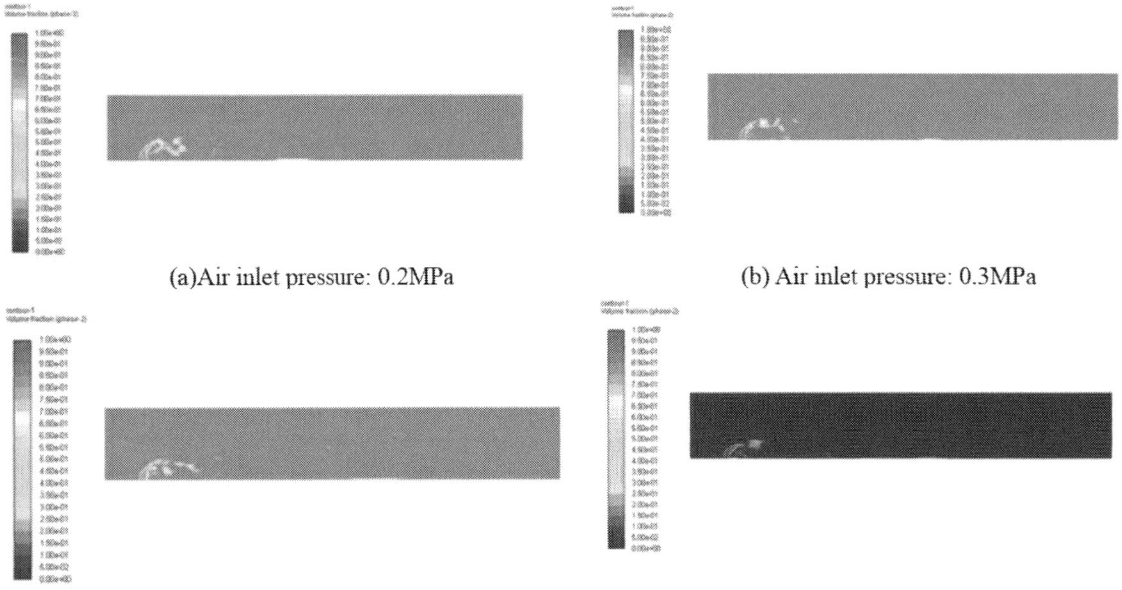

(a)Air inlet pressure: 0.2MPa (b) Air inlet pressure: 0.3MPa

(c) Air inlet pressure: 0.4MPa (d) Air inlet pressure: 0.5MPa

Fig.5. Injection depth distribution at different air inlet velocities.

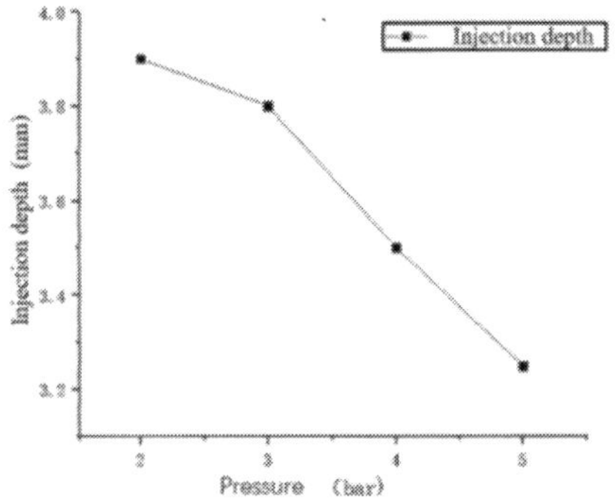

Fig.6. Influence of air inlet pressure on injection depth.

4 Experimental verification

4.1 The experimental device

The test device mainly includes a computer, mesh wire, charger, trigger, and high-speed camera. The test platform was built according to the test requirements. The AOS Technologies industrial camera was used for the high-speed camera. Under the condition of full resolution, the maximum shooting speed of this camera is more than 1000FPS. The image sensor type is CMOS, and it has a multi-head analysis system composed of two independent cameras to ensure the accuracy of the test.

4.2 The nozzle set

The corrosion inhibitor solution used in the injection-type jet device was mixed with abrasive and compressed air before the inlet end of the nozzle, the injection-type jet device was chosen to let the dust and abrasive wetted more fully and better effect of dust pollution control. The scavenger system and injection-type jet device is shown in Fig.7.

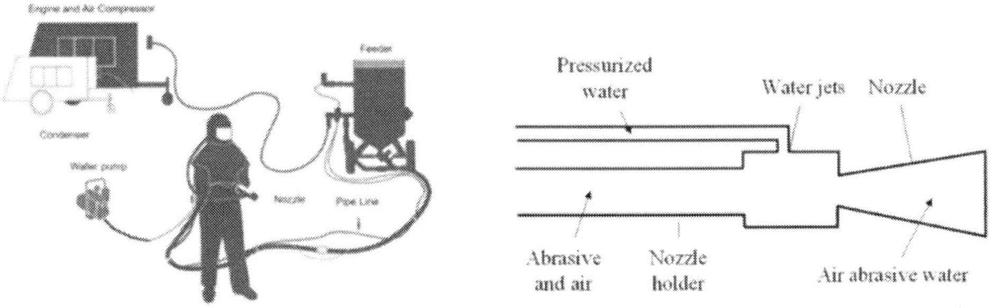

Fig.7. The scavenger system and injection-type jet device

4.3 External atomization characteristics of the nozzle

The test results of four groups, were similar, so only one of them was selected for comparison. As shown in Fig.8, with increasing the air inlet pressure, the atomization divergence area increases, and the atomized water droplets will become smaller. When the air inlet pressure is 0.2MPa, the atomization distribution at both ends of the flat nozzle is not uniform, the spray inside is mainly composed of atomized water droplets, and the dust removal efficiency is poor. When the air inlet pressure is 0.3MPa, the atomization state and dust removal effect are improved. With the constant increase of air inlet pressure, the external atomization state of the nozzle is good, and the dust removal effect is significantly improved. When the air pressure is 0.5MPa, the dust removal effect is the best, which is the parameter with higher cleaning efficiency.

(a) Air inlet pressure: 0.2MPa

(b) Air inlet pressure: 0.3MPa

(c) Air inlet pressure: 0.4MPa

(d) Air inlet pressure: 0.5MPa

Fig.8. Front atomization of the nozzle.

In the process of dry powder cleaning, when the particles impact on the glass surface, the particles will then be broken, rebound, and broken particles are the main cause of dust in the process of cleaning. Water diversion atomization suppress the broken particles, leading to achieve the purpose of dust removal.

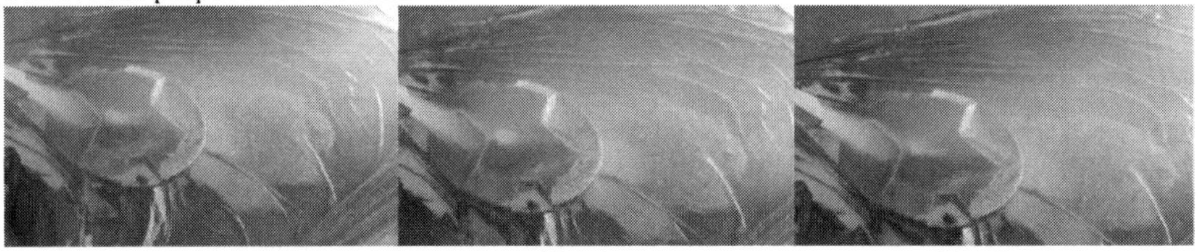

(a) 0.1s (b) 0.2s (c) 0.3s

Fig.9. The effect of no water in the cleaning process.

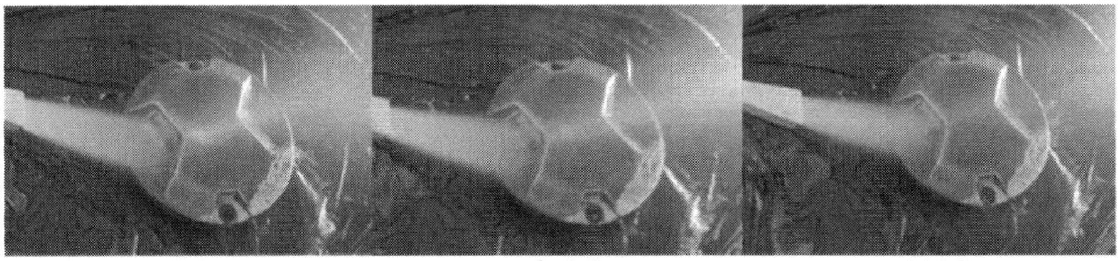

(a)0.1s (b)0.2s (c)0.3s

Fig.10. Dust removal effect in the cleaning process.

The optimized parameters for experimental observation were used to compare with the test results as shown in Fig.9. In Fig.9, when water was not added for cleaning, particles were irregularly bounced around the cleaning environment. In Fig.10, when water was added for cleaning, the bounced broken particles combined with the surrounding water mist, and the particles were wetted, thus reducing the spatter ability and removing about 90% dust. For the unbroken particles, because the particle volume was large and thus difficult to be wet by water mist, the particles splashed around the cleaning environment. According to the comparative analysis of Fig.9 and Fig.10, in the process of dry powder cleaning, the external atomization of the nozzle can effectively suppress the broken particles. When the injection pressure is 0.5MPa, the atomization dust removal effect is good. The unbroken particles will dissolve in water when washed in clean water.

5 Conclusion

1. According to the simulation analysis results of the CLSVOF method and the theory of liquid jet in transverse flow, it is concluded that the jet breaking position moves forward gradually with increasing the air inlet pressure.
2. The injection depth decreases with increasing the air inlet pressure.
3. For the external atomization state and dust removal effect of the nozzle, the experimental observation showed that when the air inlet pressure is 0.5Mpa, the atomization state and dust removal effect of the nozzle are the best, effectively reducing the harm of dust to the human body.

Acknowledgements

This work was supported by the the Foundation of External Cooperation Project in Fujian Province, China (GrantNo.2020I0027); the Natural Science Foundation of Fujian Province, China(GrantNo.2020J01289); the open fund of Fujian Provincial Key Laboratory of Functional Materials and Applications (Xiamen University of Technology), grant number fma2020002; the Program for Innovative Research Team in Science and Technology in Fujian Province University (Xiamen University of Technology).

References

1. Hu Weiwei. Development of Dry Powder Cleaning System for Vehicle Airport Runway Guide Light [D]. Xiamen University of Technology,2017.
2. Wang Xing, GE Xiaohong, Yi Mingyang, Xu Yanhui. Journal of Xiamen University of Technology,2018,26(03):20-24.
3. Gao Qingji, Wang Wenzhou, Wang Qian, Zhu Wangzhen. Journal of civil aviation university of China,2019,37(02):49-52.
4. Cheng Binbin, Yu Zhijing. Mechanism of pollution rubber layer removal of embedded airport navigation aids[J].Science Technology and Engineering,2019,19(02):86-90.
5. A. Momber. Blast Cleaning Technology[M].Springer-Verlag Berlin Heidelberg,2008.
6. Ohnesorge W. Formation of Drops by Nozzles and the Breakup of Liquid Jets [J].ZeitschriftfuerAngewandte Mathematic und Mechanik,16:355-358,1936.
7. Reitz, R. D., Atomization and Other Breakup Regimes of a Liquid Jet[J].Ph.D. thesis, Princeton University, Princeton, NJ, 1978.

Thermal properties of the three-dimensional graphene/paraffin nanocomposite phase change materials

Jing Li[1, 2*], *Yanning* Liao[1, 2], *Shaowei* Li[1, 2], *Xu* Yang[1, 2], and *Naixun* Jiao[1,2]

[1]Key Laboratory of Low-grade Energy Utilization Technologies and Systems (Chongqing University), Ministry of Education, Chongqing, 400044, China
[2]School of Energy and Power Engineering, Chongqing University. Chongqing, 400044, China

Abstract. The excellent properties of graphene phase change nanocomposite made it have potential application value in the field of heat storage materials, which was expected to achieve the integration of heat transfer and storage. In order to enhance the thermal performance of paraffin in energy storage, the structure models of n-octadecane and three kinds of graphene/n-octadecane composites were established. Molecular dynamics method was used to study the variation of thermophysical properties. It is found that the strong interaction between graphene and n-octadecane restricts the diffusion intensity of n-octadecane molecules, which reflects in the decreasing trend of the self-diffusion coefficient. In addition, the thermal conductivity of each system in the solid state is higher than that of liquid, and abruptly drops near the melting point. The thermal conductivity of the composite PCM always higher than the pure n-octadecane and increases with the amount of graphene.

1 Introduction

Nowadays, limited reserves of fossil fuels and excessive emissions of greenhouse gases make efficient use of energy a key problem to be solved. Among various forms of energy, thermal energy is widely spread in nature as in solar radiation, geothermal energy, industrial waste heat and building applications. Using phase change materials (PCMs) for thermal energy storage, can not only reduce the mismatch between supply and demand, but also improve the performance and reliability of energy distribution networks [1-2]. Phase change materials (PCMs) can solve the problem of discontinuity and instability in solar energy supply in time and realize the efficient and rational use of energy, so as to solve the energy and environment crisis [3-5].

Among numerous PCMs, paraffin (saturated hydrocarbons with C_nH_{2n+2} formula), has proved to be desirable for thermal energy storage due to its rich latent heat, low cost, thermal and chemical stability, non-toxic and non-corrosive. However, paraffin suffers from inherently low thermal conductivity which affects the rate of thermal response [6-7]. Traditional techniques to enhance the thermal conductivity of PCM is to include metallic

*Corresponding author: lj202740@cqu.edu.cn

fins and utilization of metallic foams, while their incorporation in PCM is limited in many circumstances due to their chemical activity and high density [8-11]. The discovery of graphene, consisting of a single-layer of carbon atoms that are arranged in two-dimensional honeycomb structure, has gained considerable attention around the world owing to its extraordinary physical properties [12-17]. There is still a lack of deep insights into the microscopic behaviour of melting phenomena and the enhancing mechanism of thermal transport. Molecular dynamics (MD) simulation is powerful technique to solve the above problems.

In this work, we proposed a nanostructure of the three-dimensional graphene/paraffin nanocomposites, which consist of three-dimensional porous graphene networks and energy storage material paraffin. The thermal properties of the three-dimensional nanoporous graphene/paraffin nanocomposites during phase transition are investigated systematically via MD simulations. This work will be beneficial to the development of PCM at thermal energy storage in the future.

2 Model and Method

2.1 Structure Model of Simulated systems

We performed the MD simulations both for pure paraffin and three-dimensional nanoporous graphene/paraffin nanocomposites which are applied to represent paraffin molecule. In the simulation n-octadecane is applied to represent paraffin molecule. The three-dimensional nanoporous graphene/paraffin nanocomposites, as shown in Fig. 1 (a), the graphene nanosheets are dispersed into paraffin to form the three-dimensional nanoporous graphene/paraffin nanocomposites.

2.2 Simulation method

We used MD method to calculate the established models. COMPASS force field is chosen to describe the interactions among alkane and graphene system. Algorithm is used to optimize the geometry after each structure was constructed. Then 200 ps NPT ensemble and 100 ps NVT ensemble were used to equilibrate the system.

The thermal conductivity can be calculated using Fourier's law as:

$$K = -\frac{J_z}{\left(\frac{\partial T}{\partial z}\right)_{ave}} \tag{1}$$

Where Jz is the heat flux in the z-direction, and $\left(\frac{\partial T}{\partial z}\right)_{ave}$ represents an average temperature gradient. The heat flux Jz can be calculated as:

$$J_z = \frac{1}{2A}\frac{\overline{\Delta E}}{\Delta T} \tag{2}$$

Where ΔE is the change of the kinetic energy, ΔT is the time interval, A is the cross-sectional area perpendicular to the flux direction.

3 Results and discussion

Generally, the maximum deviation of simulated density compared with theoretical value is lower than 5%, which proves that the constructed amorphous structure is effective.

3.1 Mean square displacements and Self diffusion coefficients

To rigorously characterize structural changes associated with the melting process and better demonstrate the phase change of n-octadecane and graphene/paraffin nanocomposites, we calculate mean square displacement (MSD) of atoms, which can be expressed as:

$$MSD(t)=\frac{1}{N}\left\langle\sum_{i=1}^{N}\left|r_i(t)-r_i(0)\right|^2\right\rangle \tag{3}$$

where $r_i(t)$ is the position of the atom at the given time and $r_i(0)$ represents the initial position of the atom. According to the Einstein formula, the self diffusion coefficient can be calculated as:

$$D=\frac{1}{6N}\lim_{t\to\infty}\frac{d}{dt}\sum_{i\to j}N\left\langle\left|r_i(t)-r_i(0)\right|^2\right\rangle \tag{4}$$

where D is the self diffusion coefficient and N represents the number of atoms moving diffusively in the system.

Then, the self diffusion coefficient computed from the linear regime of the MSD curve of each model at different temperatures is plotted in Fig.1 (b), the self diffusion coefficient shows obvious change at about 300K in the pure paraffin, which indicates that the phase transition of paraffin from solid to liquid is happened. For the composites, the self diffusion coefficients are all smaller than that of pure paraffin, and decrease as the mass fraction of graphene increases.

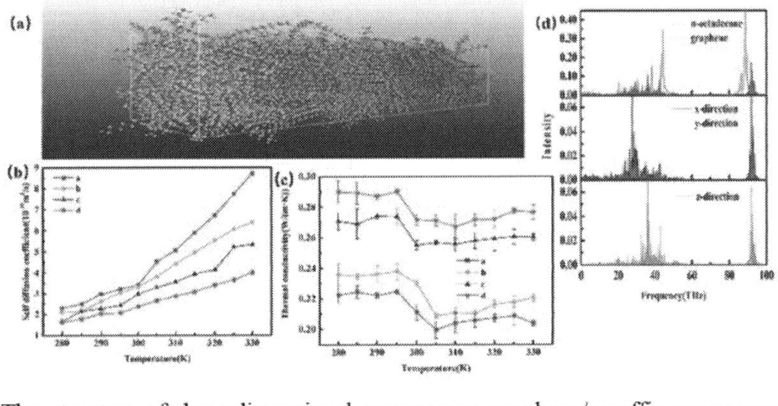

Fig. 1. (a) The structure of three-dimensional nanoporous graphene/paraffin nanocomposites. (b) Self- diffusion coefficients of pure paraffin and the composites at different temperatures. (c) Thermal conductivities of pure paraffin and the composites at different temperatures.(d) Vibrational power spectra of n-octadecane and graphene in composite system at equilibrium state of 280K.

3.2 Thermal Conductivity and Velocity auto-correlation function

The thermal conductivities of the four systems are presented in Fig. 1(c), which shows that the thermal conductivity varies with temperature and decreases sharply near the melting point. The reason for this phenomenon is that the phase change material is in a mixed state of solid and liquid near the phase transition point, the molecular movement is hindered in the solid-liquid boundary region and thermal energy is not easy to transfer between these regions. Constantly, the high thermal conductivity of graphene is more important to the enhancement of thermal conductivity of composites than the thermal resistance of solid-liquid interface, which increasing of thermal conductivity of composites obviously.

Thermal energy is essentially the energy of atomic vibration, the physical mechanism of heterogeneous coupled heat transfer in the composite system, the atomic vibration power spectra of n-octadecane and graphene were calculated respectively in this section, the expression is as follows:

$$I(\omega)=\mathrm{Re}\left[\frac{1}{2\pi}\int_0^{\infty}\langle\vec{v}(t)\cdot\vec{v}(0)\rangle e^{-iwt}dt\right] \quad (5)$$

Where $I(\omega)$ is the vibration power spectrum when the frequency is ω, $\langle\vec{v}(t)\cdot\vec{v}(0)\rangle$ is velocity auto-correlation function. The vibration power spectrum of n-octadecane and graphene in the composite system at equilibrium state 280K in Fig.1(d).

We can see that the atomic vibration power spectra of graphene and n-octadecane overlap a large part in the range of 20~43THz, while the overlap of vibration modes will help heat transfer, which is conducive to improving the heat conduction between graphene and n-octadecane molecules in the system.Furthermore,The peak of graphene at 36THz is significantly higher than others, most of which comes from the vibration mode of atoms in the normal direction of the plane, indicating that the movement of graphene atoms outside the plane has better thermal coupling with n-octadecane.

3.4 The radial distribution function

The radial distribution function (RDF) is a mathematical language to describe the internal microstructure of the matter. The radial distribution function $g_{\alpha\beta}(r)$ represents the probability of the occurrence of a β particle in a spherical shell volume r~$r+dr$ away from the reference α particle center by using the following equation:

$$g_{\alpha\beta}(r)=\frac{dN}{\rho 4\pi r^2 dr} \quad (6)$$

where ρ is density of system, N is the total number of particles in the system.

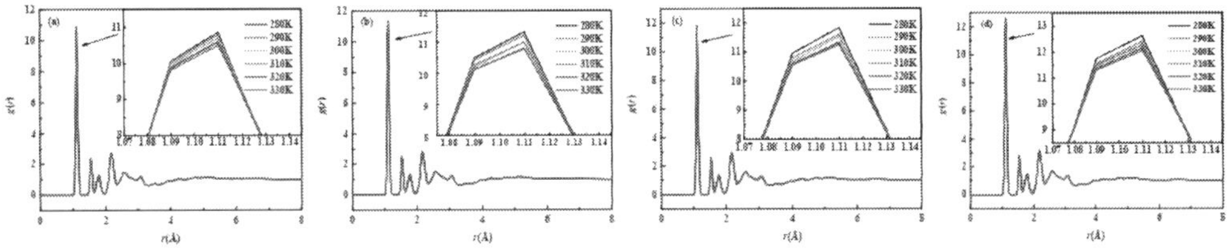

Fig. 3. Radial distribution function of n-octadecane molecules in each system at different temperatures.

Radial distribution functions of n-octadecane molecules in each system are shown in Fig.3.We can get that the radial distribution function only has peaks in the short range, showing the characteristics of short range order and long range disorder. The shape of the first peak indicates that the atomic number density within this radius is much higher than the average atomic number density of the system. By comparing the radial distribution function of each system, it was found that the peak value was higher as the mass fraction of graphene increased, which is due to the fact that the addition of graphene leads to a denser architecture, enhanced interaction with n-octadecane, and limited molecular diffusion.

4 Conclusion

Molecular dynamics simulation is applied to reveal crucial thermal properties, such as thermal conductivity ,melting temperature and heat capacity, of three-dimensional nanoporous graphene/paraffin nanocomposites during phase transition. The results could draw the following conclusion.

1.The mean square displacement and self diffusion coefficients of atoms in each system increase with the increasing temperature,and the self diffusion coefficient changes with the temperature change rate near the phase transition point

2.The thermal conductivity of each system in solid state is higher than that in liquid state, and suddenly drops near the melting point

3.With the increasing temperature, the peak value of the radial distribution function gradually decreases and the width of the peak increases slightly, indicating that the thermal motion of atoms intensifies and the order degree of the system decreases.

Acknowledgments

This work is supported by the National Natural Science Foundation of China (No.51606017), the Youth Talent Support Program of Chongqing (CQYC2021059206), Chongqing Science Fund for Distinguished Young Scholars (cstc2021jcyj-jqX0015), and the Fundamental Research Funds for the Central Universities (No.106112016CDJXY145502).

References

1. Pielichowska K, Pielichowski K. Progress in materials science, 2014, 65: 67-123.
2. Sharma A, Tyagi V V, Chen C R, et al. Renewable and Sustainable energy reviews, 2009, 13(2): 318-345.
3. Lin Y, Jia Y, Alva G, et al. Renewable & Sustainable Energy Reviews, 2018, 82.
4. Liu LK, Su D, Tang YJ, Fang GY. Renewable & Sustainable Energy Reviews, 2016, 62:305–17.
5. Tay N H S, Liu M, Belusko M, et al. Renewable & Sustainable Energy Reviews, 2016, 75.
6. Bugaje I M. International journal of energy research, 1997, 21(9): 759-766.
7. Velraj R, Seeniraj R V, Hafner B, et al. Solar energy, 1999, 65(3): 171-180.
8. Saha S K, Srinivasan K, Dutta P. Journal of Heat Transfer, 2008, 130(3): 034505.
9. Sedeh M M, Khodadadi J M. Carbon, 2013, 60: 117-128.
10. Sedeh M M, Khodadadi J M. Journal of Heat Transfer, 2014, 136(11): 112603.
11. Liu L, Su D, Tang Y, et al. Renewable and Sustainable Energy Reviews, 2016, 62: 305-317.
12. Novoselov K S, Geim A K, Morozov S V, et al. science, 2004, 306(5696): 666-669.
13. Ghosh S, Calizo I, Teweldebrhan D, et al. Applied Physics Letters, 2008, 92(15): 151911.
14. Geim A K, Novoselov K S. nature materials, 2007, 6:183-191
15. Novoselov K S, Fal V I, Colombo L, et al. nature, 2012, 490(7419): 192-200.
16. Balandin A A, Ghosh S, Bao W, et al. Nano Letters, 2008, 8(3): 902-907.
17. Chae H K,Siberio-Perez D Y, Kim J. Nature, 2004, 427(6974):523.

The interaction between *Sedum* root traits and engineered media in green roofs

Xuan Chen, *Ruifen* Liu*, and *Jun* Ma

Hubei Provincial Key Laboratory of River and Lake Ecological Restoration and Algae Utilization, Innovation Demonstration Base of Ecological Environment Geotechnical and Ecological Restoration of Rivers and Lakes, School of Civil Engineering, Architecture and Environment, Hubei University of Technology, Wuhan 430068, Hubei, China

Abstract. *Sedum sarmentosum* and *Sedum lineare*, two common plants for green roof, were planted in two engineered media meeting FLL standards at three depths. Root characteristics of root length density, root surface area density, and root volume density were quantified. And a key hydraulic property, saturated hydraulic conductivity (K_s) values of the engineered media with and without plants were measured. The results showed that engineered media type, engineered media depth, and their interaction all have a significant impact on *Sedum* root traits. For the engineered media with *Sedum*, only engineered media type and depth had a significant effect on K_s. When the type of engineered media was the same, with the same plant used, K_s of the engineered media at a depth of 14 cm was significantly smaller than K_s of any other depth. This can be attributed to the maximum level of RSD and RVD at the 14-cm depth. The reduction in K_s of the perlite-based engineered media due to plant roots is an important finding, as plant roots clog the pore spaces and lead to poor drainage and unexpected ponding in roofs. This research provides better understanding of the interaction between plant roots and engineered media in green roofs.
Keywords: Interaction; *Sedum* root traits; Engineered media; Saturated hydraulic conductivity;Green Roofs.

1 Introduction

Green roofs, as an important measure of source control and runoff control in sponge cities, have received widespread attention in recent years [1]. In modern green roof structures, plants and engineered media are two important functional layers. Plants ensure the ecological benefits of green roofs [2] and engineered media is the main site for water/solute movement. There is a close interaction between plants and engineered media.For example, plant roots can change the physical structure and hydraulic properties of engineered media, which in turn can affect the water movement within green roofs [3]. A Strong and well-developed root system would reduce the rainwater retention effect of green roofs due to preferential flow related to root-induced structure[4-5].The type and depth of engineered media are important factors affecting the growth of plants[6-7], which are closely related to

*Corresponding author: ruifen1986@aliyun.com

plant above-ground biomass and survival rate[8]. It was found that plants grown in shallower substrate had better biomass accumulation in green roofs[9]. However, little is known about how engineered media affects plant root traits. As a key hydraulic property of engineered media, saturated hydraulic conductivity (K_s) is generally quantified without considering the role of plant roots [10]. The change in K_s due to the presence of plant roots remains largely unknown. In this study, *Sedum sarmentosum* and *Sedum lineare*, which are two commonly used plants in green roofs, were planted in two engineered media meeting FLL standards [11] at three depths. Root traits of length density, surface area density, and volume density were measured. K_s values of the engineered media with and without plants were also determined. The purposes of this study were to quantify the effects of engineered media type and depth on *Sedum* root traits and to compare K_s values of engineered media varying in type and depth with different root traits.

2 Materials and methods

2.1 *Sedum* type and engineered media

Two engineered media were developed by mixing aggregate and manure at a volume percentage of 9:1 (Table 1). The first engineered media mix used perlite aggregate and the second used vermiculite aggregate. Both engineered media were developed according to FLL standards, and their basic characteristics are listed in Table 1.

Table 1. Characteristics of the engineered media [mean (standard error)].

Characteristics	Perlite-based mix	Vermiculite-based mix
Components(% by volume)	90 % perlite(<6 mm) 10 % chicken manure	90 % vermiculite(<5 mm) 10 % chicken manure
Dry bulk density(g/cm)	0.21(0.01)	0.34(0.01)
Total porosity(%)	91.40(0.01)	78.80(0.02)
Maximum water-holding capacity(%)	36.65 (1.33)	64.05(1.55)
Organic matter content(g/kg)	31.15(2.72)	38.64(2.60)
K_s(cm/min)	54.45(0.19)	18.48(1.39)

Sedum sarmentosum and *Sedum lineare* were chosen for this study for their wide application in extensive green roofs [12-14]. Considering that *Sedum* plants are shallow-rooted species with weak edge effects [15], acrylic cylinders configured with perforated bases for drainage, and of 10-cm diameter and 15-cm height, were used for indoor small-scale cultivation. Plant cultivation involved 12 treatments, including two types of plants (*Sedum sarmentosum* and *Sedum lineare*), two types of engineered media (perlite-based mix and vermiculite-based mix), and three engineered media depths (6, 10, 14 cm). Each treatment had three replicates, making 36 samples in total.

In each sample, *Sedum* cuttings were planted at a spacing of 2 cm × 2 cm. Thereafter, samples were put into an artificial climate chamber (Shanghai Yiheng MGC400H) and subjected to a typical climate in Wuhan City. The temperature was maintained at 25 °C, relative humidity was set between 40 % and 50 % and a daily photoperiod of 10:14 h (L:D) was scheduled [16]. In addition, 85 mL water was regularly applied to each sample every 3 days. After 103 days, samples were taken out for the measurements of root traits and K_s.

2.2 Measurement of *Sedum* root traits

Before measurement, the roots were cleaned and plucked using brushes and forceps to minimize crossover and overlap. Root images were obtained by a German PMT-RTP-A3 root system scanner (the pixel setting was 600 dpi.). Scanned images were then analyzed by the Root Analysis system to obtain various root traits including root length density (RLD, mm/cm^3), root surface area density (RSD, mm^2/cm^3), and root volume density (RVD, mm^3/cm^3).

2.3 Measurement of K_s

The K_s values were measured according to the falling-head method specified by the FLL standard [11], before being taken out of plants in samples for root trait measurement. The detailed steps were as follows. After cutting off the above-ground part of *Sedum*, samples with various depths (H, cm) were submerged in water for at least 24 h and then left to drain for up to 2 h. A two-pronged metal ring was placed on the tops of the samples. Water was continually added to the tops of the samples until a constant water level was maintained on the tops and water drained freely out of the perforated bottoms. K_s was then calculated based on the time (t, s) taken for the water level to drop from the 45 mm-prong to the 35 mm-prong: K_s (cm/s) = H/(H·t + 0.4t).

2.4 Statistical analysis

All data analyses were performed in IBM SPSS Statistics 26.0. Effects of engineered media type and depth on *Sedum* root traits and K_s comparison of engineered media with various root traits were analyzed using a two-way analysis of variance (ANOVA). Statistical differences between compositions were assessed at the $P<0.05$ level of significance by t-test (for a comparison between two) or LSD post hoc test (for a comparison between three). Data were checked for normality and were transformed appropriately (using the Box-Cox transformations) where necessary before analysis. All data presented in figures and tables were non-transformed.

3 Results and discussion

3.1 Effects of engineered media type and depth on *Sedum* root traits

For the RLD of *Sedum sarmentosum*, engineered media type (F=259.97, $P<0.01$), engineered media depth (F=366.39, $P<0.01$), and their interaction (F=150.10, $P<0.01$) all showed significance, indicating that the two factors jointly influenced the changes in the RLD [Fig. 1(a)]. The RLD of 9.60 mm/cm^3 in the 6-cm vermiculite-based engineered media was the largest. Similar to *Sedum sarmentosum*, engineered media type (F=477.10, $P<0.01$), engineered media depth (F=1672.86, $P<0.01$), and their interaction (F=258.33, $P<0.01$) resulted in significant differences in the RLD of *Sedum lineare*. The largest RLD was recorded in 6-cm vermiculite mixed engineered media, at 10.52 mm/cm^3. The RLD is the length of the root system contained per unit volume of a growing media and reflects the degree of root extension and interweaving within the media [17]. It is generally accepted in soil science that the deeper the soil, the more extensive the plant roots [18,19]. This is because deeper soils can not only provide more water/fertilizer for plant growth but also maintain a more stable temperature beneficial for plant root growth [20]. However, in this study, these media depth-related benefits for plant growth were rather limited by a narrow media depth range of 6 cm to 14 cm. As a result, the RLD was greatest at a minimum media depth of 6 cm. The results of Lu et al [8], who studied the impacts of media depth on the morphology of *Sedum lineare* in a green roof system, also showed that the growth of root length in a shallow media of 4 cm was greater than the growth in a 10 cm media.

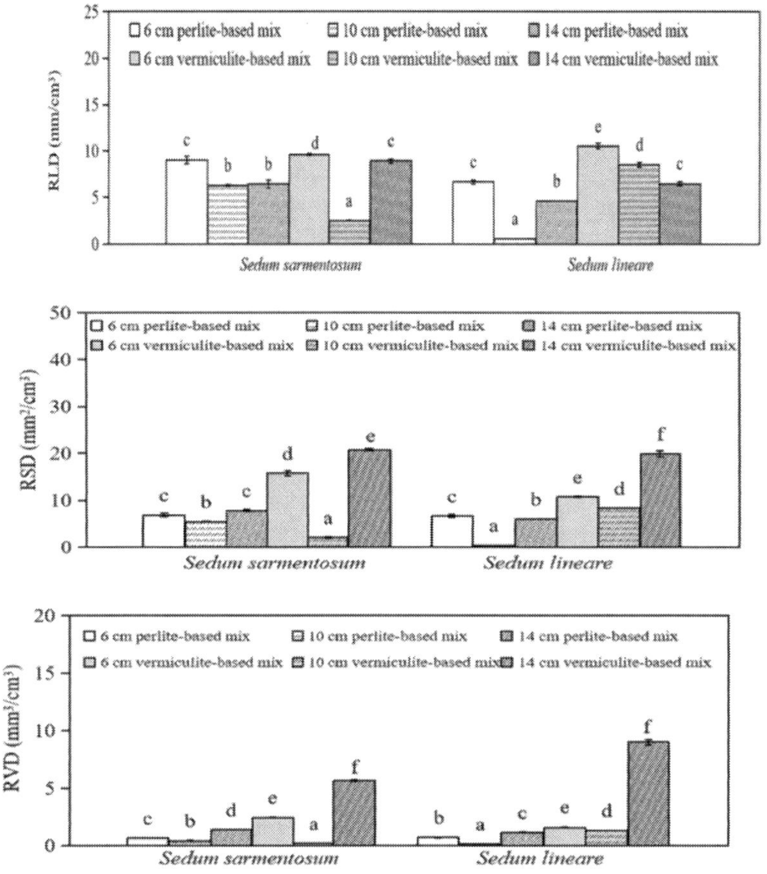

Fig. 1. Root length density (a), root surface area density (b), and root volume density (c) of *Sedum* plants within each treatment. Where there is a significant interaction, lowercase letters denote significant differences between all means within the same plant.

With regards to the RSD of *Sedum sarmentosum*, engineered media type (F=1172.44, P<0.01), engineered media depth (F=1206.85, P<0.01), and their interaction (F=742.56, P<0.01) all showed significance, indicating that the two factors both influenced the changes in the RSD together [Fig. 1(b)]. The RSD was the largest (20.76 cm^2/cm^3) in the 14-cm vermiculite-based engineered media and was at least 1.31 times greater than any other group. As with *Sedum sarmentosum*, engineered media type (F=925.65, P<0.01), engineered media depth (F=2780.13, P<0.01), and their interaction (F=297.69, P<0.01) resulted in significant differences in the RSD of *Sedum lineare*. The RSD was highest (19.91 cm^2/cm^3) and at least 1.84 times greater than any other group in the 14-cm vermiculite-based engineered media.

For the RVD of *Sedum sarmentosum*, engineered media type (F=9635.87, P<0.01), engineered media depth (F=8910.10, P<0.01), and their interaction (F=4375.87, P<0.01) all exhibited significance, indicating that the two factors jointly influenced RVD changes [Fig. 1(c)]. The RVD was the largest (5.66 cm^3/cm^3) in the 14-cm vermiculite-based engineered media and was at least 2.31 times greater than any other group. Comparably to *Sedum sarmentosum*, engineered media type (F=19549.06, P<0.01), engineered media depth (F=10016.15, P<0.01), and their interaction (F=1889.29, P<0.01) all resulted in significant differences in the RVD of *Sedum lineare*. The maximum level of RVD (8.99 cm^3/cm^3) was recorded in the 14-cm vermiculite-based engineered media, which was at least 5.62 times greater than any other group. The RSD and RVD can indicate the contact area between the root system and the engineered media [21], and together with RLD, can reflect the root traits. The 14-cm vermiculite-based engineered media had the largest RSD and RVD but a relatively smaller RLD, suggesting that media depth-related benefits for plant roots are embodied by RSD and RVD. In this case, the *Sedum* root system limited RLD development but promoted RSD and RVD development to obtain more water and nutrients in the shallow depth environment.

3.2 K_s comparison of engineered media with various root traits

For the engineered media with *Sedum*, the engineered media type and depth had a significant effect on K_s, but their interaction had no significant effect (Figure 2). The role of media type and depth on K_s is expected, as the media type and depth had an impact on the root traits (Section 3.1). The fact that the interaction of media type and depth is significant for *Sedum* root traits but not for K_s, indicates that this interaction is not big enough to change K_s. As shown in Figure 2, when the depth of the engineered media was the same, with the same plant, K_s in the perlite-based engineered media was significantly smaller than that in the vermiculite-based engineered media. K_s in plants in the perlite-based engineered media ranged from 0.568-2.641 cm/min with a mean value of 1.455 cm/min, while K_s in the plants in the vermiculite-based engineered media ranged from 12.77-19.19 cm/min with a mean value of 15.39 cm/min. The latter result was therefore about 6.42-22.72 times higher than the former. When the type of engineered media was the same, with the same plant, K_s of the engineered media at a depth of 14 cm was significantly smaller than K_s of any other depth. This can be attributed to the maximum level of RSD and RVD recorded in the 14-cm depth (Section 3.1). A potential negative correlation exists between K_s and RSD/RVD.

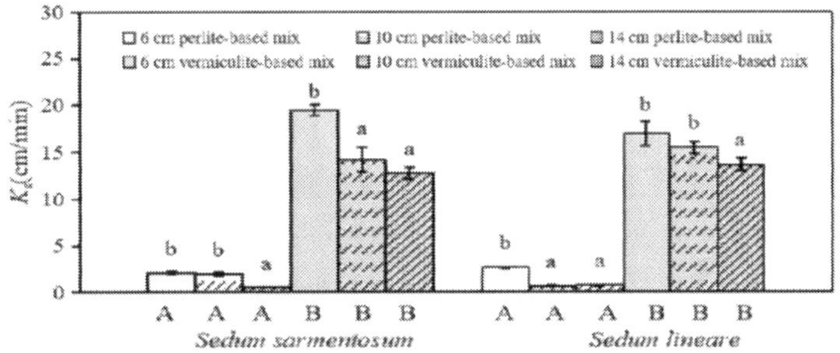

Fig. 2. Saturated hydraulic conductivity of engineered media with *Sedum*. Where there is no significant interaction, uppercase letters denote significant differences between engineered media types and lowercase letters denote significant differences between engineered media depth.

Compared with perlite-based engineered media without plants (Table 1), K_s of perlite-based engineered media with plants were significantly smaller and had a greater variation. It is well known that plant roots can significantly reduce K_s by up to 143% [22]. Our study found similar results as the presence of roots in the perlite-based engineered media resulted in a reduction in K_s of over 95%. This may be due to the hard texture of perlite particles. Without changing the existing positions of perlite particles, *Sedum* roots can only grow along the curved interparticle paths [23], which reduces the total porosity and decreases the K_s of engineered media [24]. The reduction in K_s of the perlite-based engineered media due to plant roots is noteworthy, as plant roots would clog the pore spaces and lead to poor drainage and unexpected ponding in roofs. As for the vermiculite-based engineered media, there was no significant difference in K_s between the plant group and the no-plant group. The above results indicate that depending on engineered media type, changes in K_s due to plants roots can vary remarkably.

4 Conclusions

Engineered media type, engineered media depth, and their interaction all showed significance on *Sedum* root traits. For the engineered media with *Sedum*, only engineered media type and depth had a significant effect on K_s, indicating that this interaction could only change the *Sedum* root traits, but it is not big enough to change K_s. When the depth of the engineered media was the same, with the same plant, K_s in the perlite-based engineered media was significantly smaller than that in the vermiculite-based engineered media. When the type of the engineered media was the same, with the same plant, K_s of the engineered media at a depth of 14 cm was significantly smaller than K_s of any other depth. This can be attributed to the maximum value of RSD and RVD in the 14-cm depth. A potential negative correlation exists between K_s and RSD/RVD. Compared with perlite-based engineered media without plants, K_s of perlite-based engineered media with plants were significantly smaller. The presence of roots in the perlite-based engineered media resulted in a reduction in K_s of over 95%. As for the vermiculite-based engineered media, there was no significant difference in K_s between the plant group and the no-plant group. The above results indicate that depending on engineered media type, changes in K_s due to plants roots can vary remarkably.

Future work should focus on the study of other hydraulic properties of engineered media induced by roots in green roofs. This will better understand the water transport process of green roofs in sponge city construction.

Acknowledgements

The work is supported by the National Natural Science Foundation of China (51909081 and 5217090897).

References

1. M. Xiao, Y. Lin, J. Han, Renewable and Sustainable Energy Reviews, 40 (2014)
2. K. Vijayaraghavan, Renewable and Sustainable Energy Reviews, 57 (2016)
3. A. Skorobogatov, J He, A Chu, Sci Total Environ, 715 (2020)
4. Z Zhang, C. Szota, T.D. Fletcher, Nicholas S.G. Williams, J. Werdin, C. Farrell, Sci Total Environ,625 (2018)
5. Y.C. Hu ,H.P. Qin , Z.X.Lin, Journal of Shenzhen University(Science & Engineering),37 (2020)
6. C.Brown, J.Lundholm, Landscape & Urban Planning, 143 (2015)
7. J.Chenot, E.Gaget, C.Moinardeau, Multidisciplinary Digital Publishing Institute, 9 (2017)
8. J. Lu, J.G. Yuan, J.Z. Yang,Ecological Engineering, 74 (2015)
9. J. Lu, J.G. Yuan, J.Z. Yang,Urban Forestry & Urban Greening, 13 (2014)
10. S.L. Peng, W.H. You, H.T. Shen, Transactions of the Chinese Society of Agricultural Engineering, 26 (2010)

11. FLL. Duidelines for the Planning, Construction and Execution of green-roof site(Germany Troisdorf: Forschungsgesellschaft Landschaftsentwicklung Landschaftsbau e.V, 2008)
12. T. Luo, W.N. Xu, H. Cheng, Environmental Engineering, 38 (2020)
13. S.X. Zhang, S.H. Zhang, Y. Zhang, Environmental Science, 40 (2019)
14. T. Li, Y.L. Chen, J.Q. Gu, Journal of Tongji University(Natural Science), 43 (2015)
15. Y. Dusza, S. Barot, Y. Kraepiel, Ecol Evol, 7 (2017)
16. M.H. Yu, P. Zhao, C.P. Ceng, Guihaia, 32 (2012)
17. J.X. Li, B.H. He, Y.Chen, Transactions of the Chinese Society of Agricultural Engineering, 29 (2013)
18. S. Tron , G. Bodner, F. Laio, Ecol Modell, 312 (2015)
19. K.L. Getter , D.B. Rowe, Urban Ecosystems, 11 (2008)
20. L.H. Comas, S.R. Becker, V. Cruz, Frontiers in Plantence, 4 (2013)
21. J.X. Li, B.H. He, Y.Chen, Acta Ecologica Sinica, 33 (2013)
22. J. Lu, Q. Zhang, A.D. Werner, Journal of hydrology (Amsterdam), 589 (2020)
23. A.G. Bengough , B.M. Mckenzie , P.D. Hallett, Journal of Experimental Botany, 62 (2011)
24. N. Koebernick, K.R. Daly, S.D. Keyes, New Phytologist, 216 (2017)

Study on low-temperature treatment of nitrogen-containing wastewater by anaerobic biological filter with alkali-treated loofah-based filler

Jiaxu Guang[1,2], *Jiayi* Lu[1,2], and *Qing* Ting[1,2,*]

[1] National Environmental Protection Textile Industry Pollution Control Engineering Technology Center,Shanghai,China
[2]College of Environmental Science and Engineering, Donghua University,Shanghai,China

Abstract. In this study, a new type of anaerobic biological filter (ABF) packed with alkali-treated loofah filler and a filler-free suspension growth control group (SBR) were used to cultivate the inoculated activated sludge to improve the activity and low temperature tolerance of the enriched anaerobic ammonia oxidation bacteria in the reactor. The results show that ABF and SBR were acclimated and started quickly, and the total nitrogen removal rate was maintained at 70.9% and 63.7%, respectively. Under the low temperature operating conditions, the total nitrogen removal rate of ABF maintained 59.9%, which was 16.5% higher than that of SBR. The detected EPS in the biomass of ABF and SBR increased by 100% and 272%, respectively, under low temperatures (15℃). Metagenomic testing revealed that the anammox bacteria (*Candidatus Brocadiae*) were the dominant functional populations both in ABF and SBR systems, and their relative abundances were 22.5% and 11.38%, respectively. In addition, it is found for the first time that anammox bacteria and denitrifying bacteria were the main contributors to EPS synthesis, which is important to the stable operation of the reactor at low temperature.
Keywords: Low temperature operation; Anammox; Alkali-treated loofah packing; Biological filter; Extracellular polymer; Metagenomic.

1 Introduction

In recent years, the process of Anaerobic Ammonia Oxidation (ANAMMOX) has attracted extensive attention and has been applied in the full-scale urban and industrial wastewater treatment plants[1], with its advantages of low requirement for aeration and carbon sources, and yields less wasted sludge[2].The enrichment of Anammox bacteria (AAOB) is the key to start the anammox process and improve their tolerance to cold environment rapidly. Studies have shown that the extracellular polymer secreted by AAOB may enhance their attachment to the biological filler and thus effectively alleviate the cell erosion phenomenon inside the

*Corresponding author: tq2004@dhu.edu.cn

fillers, which could increase the mechanical and ecological stability of the system[3].

Nowadays, most of the carrier fillers used in the process are natural or synthetic polymer fillers with simple structures and difficult for biofilm development. Loofah has the potential as good biological carrier filler [4] with various advantages, such as low cost, porous structure with high specific surface area, non-toxic and high affinity for microorganisms growth[5].The alkali-treated loofah material reduces the hydroxyl groups on the fiber surface and improves the fiber's microbial corrosion resistance and the service life of the filler.

Therefore, in this experiment, a new type of alkali-treated loofah-based filler biological filter was used to enhance the enrichment of the bacterial population in the reactor by adding porous structure fillers, and to improve the biological denitrification efficiency of the reactor. The research results can provide new ideas for maintaining the nitrogen removal efficiency of the anaerobic ammonia oxidation process under low temperature operation and improving the tolerance of the bacterial flora to low temperature environments.

2 Materials and Methods

2.1 Sludge and Experimental Influent

The inoculated sludge was sludge obtained from the second settling tank of Shanghai Songjiang Sewage Treatment Plant, with a sludge concentration of 10.08g/L. The feeds applied in the experiments were the synthetic textile industrial wastewaters. During the overall denitrification, the formula of the wastewaters are: 120mg/L of NH_4^+ and 140mg/L of NO_2^-, 100mg/L of COD, 300mg/L of $MgCl_2 \cdot 6H_2O$, 137mg/L $CaCl_2$ and 30mg/L KH_2PO_4 were used as the nutrients of the reactor, and $NaHCO_3$ was used to maintain the pH of the influent water within the range of 7.5±0.5.

2.2 Experimental Device Parameters and Operation

The commercially available loofah was firstly cut into cylindrical slices with a thickness of 1 cm and soaked in 80% NaOH solution for 12 hours and dried to prepare the new type of loofah-based filler. The experimental anaerobic biological filter (ABF) was packed with the above-mentioned alkali-treated loofah-based fillers. In addition, the control group (SBR) was used for comparison. The daily processing capacity of both devices were 30L. The experiment was divided into two phases: (1), the 35℃ stable start-up operation phase and (2), the 15℃ low temperature tolerance test phase. In the first stage, black opaque thermal insulation foam material were used to wrap the experimental device, and the two sets of biological filters were maintained at 35°C for 56 days. In the second phase, the reactor temperatures were reduced gradually. The temperature controlling system kept the two biofilters operated at a room temperature of 15°C for 42 days.

2.3 Analysis Method

The water samples were filtered through a 0.45μm membrane before monitoring. The alkaline persulfate digestion and ultraviolet spectrophotometry methods were used to determine the TN, NH_4^+-N, NO_2^--N and NO_3^--N of the wastewater samples.

The sludge EPS was extracted by physical centrifugation method [6]. The polysaccharides and protein in the biomass were determined with anthrone-sulfuric acid assay [7] and bicinchoninic acid(BCA) method[8], respectively.

After the experimental low-temperature phase is over, the biomass in the biological filter were washed with physiological saline before dried with centrifugation and sent to the bio-company (Shanghai Meiji Biotechnology Co., Ltd) for metagenomic analysis.

3 Results and Discussion

3.1 Removal Effect of Nitrogenous Pollutants

Start two sets of reactors, with the increase of operating time, the concentration of effluent NH_4^+ and NO_2^- decreased gradually, and the denitrification efficiency increased gradually,

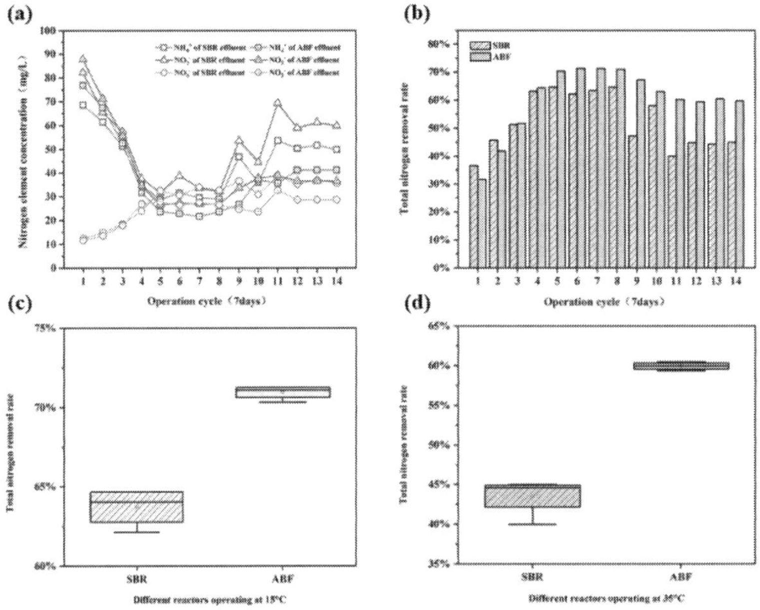

Fig. 1. Nitrogen removal performance of SBR reactor and ABF reactor after 14 weeks of operation: **(a)** nitrogen content of in and out water, **(b)** total nitrogen removal rate, **(c)** total nitrogen removal efficiency at 15℃, **(d)** total nitrogen removal efficiency at 35℃.

as shown in Figure 1(a). In the fourth week, the various indicators of the reactor stabilized. The denitrification efficiency of the SBR reactor and the ABF reactor were stabilized and reached average 63.7% and 70.9%, respectively from the fourth week to the eighth week. Tthe denitrification rates were 97.6. mg/(L·d) and 90.9mg/(L·d), respectively, as shown in Figure 1(b). The total nitrogen removal rate was about 60%, which was nearly 15% higher, compared with the SBR reactor. After the temperature control system was removed, as the temperature decreased, the removal rate of pollutants decreased significantly with the anammox bacteria inactivated gradually. The reason for this situation may be the sudden drop in temperature and the communities in the biomass shifted for re-adaptation as the environmental temperature decreased [9]. During the fifth to eighth week in the first stage and tenth to fourteenth week in the second stage, when the reactor was operating stably at different temperatures, the nitrogenous pollutants removal in ABF were much better than

SBR as shown in Figure 1(c) and Figure 1(d).

3.2 EPS Analysis

The composition of EPS generally varies with different bacterial communities, and micro-ecological environment composed of water (97%), extracellular polysaccharides, extracellular proteins, etc. [10]. The change of EPS composition were tracked in the operation of the two reactors. It was found that the main organic components of the biomass selected in the experiment were protein and polysaccharides, and the content of protein was much higher than that of polysaccharides, as shown in Table 1. The EPS inoculated with activated sludge was mainly composed of fixed extracellular polymer, which accounted for 84.5% and 86.1% of the total extracellular polymer in SBR and ABF reactors[11].

Table 1. EPS content in activated sludge inoculated in the reactor.

Reactor	EPS category	S-EPS	LB-EPS	TB-EPS
SBR	Polysaccharide (PS)	1.58	1.74	3.95
	Protein (PN)	5.71	11.95	21.95
	Total amount of EPS	7.28	13.69	25.90
ABF	Polysaccharide (PS)	1.22	1.11	2.94
	Protein (PN)	3.28	8.51	15.40
	Total amount of EPS	4.51	9.62	18.34

The polysaccharide (PS) and protein (PN) EPS content tests were performed on the SBR reactor and the ABF reactor, respectively, as shown in Figure 2.

After the two reactors were operated for 50 days under temperature-controlled conditions, the soluble EPS content increased significantly. The activated sludge in the SBR reactor gradually increased its sedimentation and formed granular sludge, while the ABF

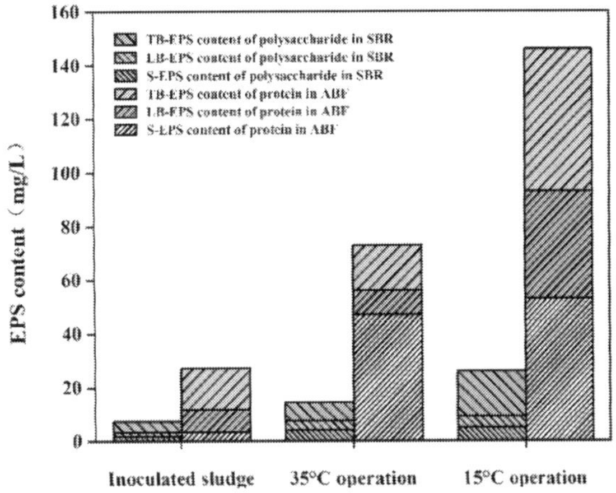

Fig. 2. Analysis of EPS content in the reactor at different operating stages.

reactor gradually formed a biofilm attached to the filler. The sludge form and the function of the biological community of the two gradually changed, so SBR and ABF reactors nitrogen removal performances were steadily improved with the stable operation of the reactor, as shown in Figure 1.

With the advancement of the reactor operation process, the content of fixed and soluble polysaccharide EPS in the SBR reactor gradually increased. After the temperature control system was removed, the composition of soluble EPS remained almost unchanged, and the content of fixed EPS increased greatly. After 50 days of stable operation of the ABF reactor, the total amount of EPS increased significantly. Among them, the change of fixed EPS or TB-EPS was relatively small. After the temperature control system had been removed, the content of soluble EPS remained almost unchanged, and the content of fixed EPS has increased significantly (with an increase of 272%).

3.3 Analysis of Bacterial Colony Composition and EPS Synthesis Pathways

Metagenomic sequencing and bioinformatics analysis were performed on ABF and SBR. Both reactors were detected for their biomass bacterial composition, which was dominated by "*Candidatus_Brocadiae*". Among them, the relative abundance of anammox in ABF and SBR were 22.5% and 11.38% respectively, as shown in Figure 3.

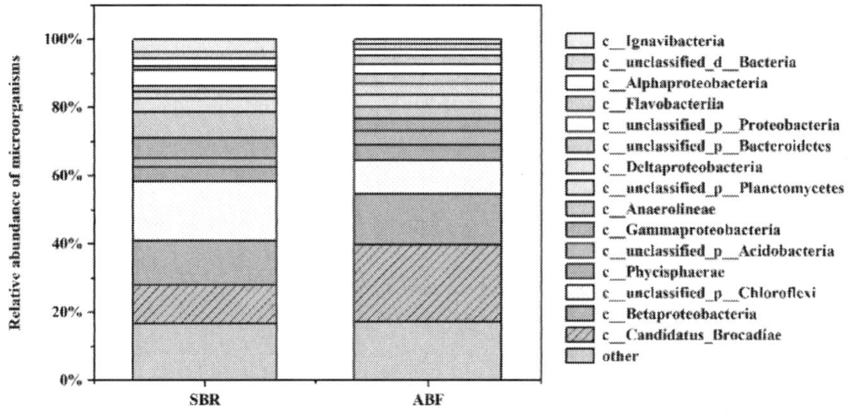

Fig. 3. The level composition of the microbial genus in the biofilter with an abundance of more than 1%.

In addition, the accumulation of EPS could assist in developping cold resistance and buffering capacity to unfavorable growth factors such as low temperature and toxicity, which is the reason why the addition of fillers under low temperature conditions helps the bacterial community to grow better.

As shown in Figure 4, the anammox bacteria and denitrifying bacteria were the main

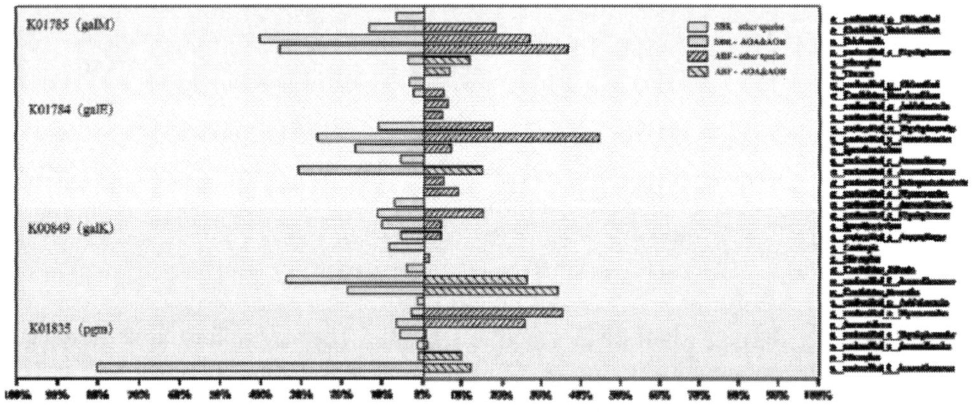

Fig. 4. The relative abundance of the four types of genes coding for EPS synthesis by different community members in the ABF and SBR operated under 15℃ condition.

functional populations that synthesize EPS. Among them, the genus of *unclassified_f__Anaerolineaceae, Nitrospira, Candidatus_Brocadia, Candidatus_Jettenia, Thauera* were the five typical ammonia oxidizing archaea (AOA) and ammonia oxidizing bacteria (AOB), which contributed high fraction of genes coding for different pathwaysinEPS synthesis.

Both reactors were dected for typical EPS synthesis pathways via the GALM, GALE, GALK and PGM. As shown in Figure 4, in the two EPS synthesis pathways of GALM and GALK, the AOA and AOB species in the ABF reactor accounted for 14.8% and 5% higher, respectively, than those in the SBR reactor. In the gene expression process of PGM and GALM, the species of AOA and AOB involved in EPS synthesis in100% more than that in SBR. The above data showed that compared to the suspension culture of the SBR reactor, the ABF reactor with loofah as the filler not only promoted the growth and enrichment of the bacterial flora, but also potentially improved the adhesion efficiency of the ammonia oxidizing bacteria with the synthesied EPS. On the basis of the above experiments, a new method is proposed for the stable operation of the low-temperature environment bioreactor.

4 Conclusions

Using ABF and SBR to treat nitrogen-containing wastewater, the reactor can be started quickly when operating at 35°C, and the average total nitrogen removal rate after 56 days of stable operation of the reactor were 70.9% and 63.7%, respectively. In a low temperature environment(15℃), the accumulation of EPS in ABF and SBR was positively correlated with the total nitrogen removal rate and low temperature tolerance of the reactor.

The relative abundance of dominant functional bacteria (*Candidatus_Brocadiae*) in ABF and SBR accounted for 22.5% and 11.38%, respectively, which is an important biological factor for ABF with the low temperature tolerance.

AOA&AOB colonies account for a relatively high proportion of the synthesis of EPS, which provides an energy basis for the denitrification of the biological community under low temperature conditions.

Acknowledgement

Carbon source regulation mediated by phosphorus recovery-low temperature resistance mechanism of biofilm nitrogen and phosphorus removal system (17K11319)

References

1. K.B. Alpaslan, D. Duygu, S. Neslihan, Water Sci. Technol, **78** (2018)
2. B. Kartal, W.J. Maalcke, N.M.D. Almeida, Nature, 479 (2011)
3. E. Steffen, F.M. Terence, R. Kristina, Water Sci. Technol, 74 (2016)
4. J. Shen, Y.M. Xie, X. Huang, J. Mech. Behav. Biomed. Mater, 15 (2012)
5. J. Shen, Y.M. Xie, X. Huang, Int.J. Impact Eng, 57 (2013)
6. T.T. More, J.S.S. Yadav, S. Yan, J. Environ. Manage, 144 (2014)
7. F. Leng, S. Sun, Y. Jing, Bulg. Chem. Commun, 48 (2016)
8. K. Pavel, K. Tatyana, B. Maria, Int. J. Pharm, 599 (2021)
9. T. Lotti, R. Kleerebezem, J.M. Abelleira-Pereira, Water Res, 81 (2015)
10. C. Pellicer-Nàcher, C. Domingo-Félez, A.G. Mutlu, Water Res, 47 (2013)
11. I. Sharif, E. Maria, O.J. A, Environ. Technol, 38 (2017)

Experimental study on the crack resistance of steel-nanometre hybrid fibre concrete

Hailin Wu[1,*], *Zijia* Mi[2], and *Ziqiang* Pei[3]

[1]College of Hydraulic and Environmental Engineering, China Three Gorges University,443002 Yichang,China
[2]College of Hydraulic and Environmental Engineering, China Three Gorges University,443002 Yichang,China
[3]Yellow River Wanjiazhai Water Control Project Co., Ltd,030009 Taiyuan,China

Abstract. To investigate the effects of steel fibre and nanometre fibre on the cracking of reinforced concrete, the cracking resistance of steel fibre reinforced concrete, nanometre fibre reinforced concrete, and hybrid fibre reinforced concrete with different doping levels were studied in comparison with the baseline concrete specimens by steel-Nano hybrid fibre reinforced concrete axial tensile tests. The results show that when the steel fibre admixture is 1.5% and the nanometre fibre admixture is 0.05%, the initial cracking load of reinforced steel-Nano hybrid fibre axial tensile specimens is enhanced the most, at this time, compared with the initial cracking load of plain concrete specimens by 58.9%, at the same steel stress level, when the steel fibre admixture is 1.5% and the nanometre fibre admixture is 0.10%, it is most effective for crack control.
Keywords: Reinforced concrete; Nanometre fibre; Steel fibre; Axial tensile test.

1 Introduction

As we all know, the problem of concrete cracks is inevitable. Nowadays, even many components work with cracks, such as the current widely used steel-lined reinforced concrete pipes in hydropower station projects, to give full play to their mechanical properties, the pipes are generally working with joints in the state, but the existence and development of cracks will cause a very serious problem, namely, the internal reinforcement corrosion occurs. The corrosion of the reinforcement will inevitably affect the load-bearing capacity and durability of the concrete material, and even endanger the overall safety of the building structure, so it is urgent to solve the cracking problem of concrete.

Blending fibres in concrete can well compensate for the low tensile strength, brittleness, and poor toughness of concrete. It is proved [9-12] that there are certain limitations to improving the performance of concrete by incorporating only one type of fibre, and incorporating two or more fibres into concrete can produce synergistic effects to give full play to their advantages and improve the mechanical properties of concrete, so hybrid fibre

* Corresponding author:82667105@qq.com

concrete should be born. In this paper, from the perspective of crack control, steel-Nano hybrid fibre reinforced concrete specimens are designed and tested in axial tension to study the effects of steel fibre admixture and Nano fibre admixture on the initial crack load, crack morphology, crack spacing and crack width of the specimens.

2 Test methodologies

2.1 Test materials

The strength of the reinforced concrete pressure pipeline is generally C20-C30, and the strength of the concrete matrix used in this test is C25, the coarse aggregate is crushed stone with a grain size of 5 mm-20 mm, the fine aggregate is high-quality river sand, the water reducing agent is polycarboxylic acid master batch with a solid content of 40±1%, the water used is tap water. The water used is tap water. The specific parameters of the selected steel fibre and nanometre fibre are listed below.

Table 1. Physical properties of steel fibre.

Fibre Name	Aspect ratio L/D	Types	Density (g/m^3)	Tensile strength (MPa)	Elongation (%)
Steel fibre	80	Shear corrugated type	7.8	≥600	-

Table 2. Physical properties of nanometre fibre.

Fibre Name	Length (μm)	Widths (nm)	Morphological ratio	Exterior appearance	Concentration (%)	Tensile strength (MPa)
Nanometre fibre	>1	20-50	>30	Transparent gel	4.85	222-233

2.2 Experimental program

The cross-sectional dimensions of the steel-Nano hybrid fibre axial tensile specimen are 100 mm × 100 mm and the length is 700 mm, and the force transmission of the specimen is using direct pulling of the main reinforcement.

The volume rates of steel fibre determined for this test were 0.5%, 1.0%, and 1.5%, the nanometre fibre was dosed at 0.05%, 0.10%, 0.15%, and 0.20% of the cement mass, in addition to control groups designed with steel fibre alone, nanometre fibre alone and no fibres. Twenty specimens were designed and fabricated and the basic parameters of the specimens are shown in Table 3 below. Note on specimen numbering: S and N represent the volume rate of steel fibre and nanometre fibre doping, respectively.

2.3 Effect of mixed fibres on the initial cracking load

In this test group, the test machine load corresponding to the sudden stress change in the steel reinforcement was used as the initial cracking load of the specimen by real-time monitoring, and then the results of the initial cracking load of the specimen were divided into three groups of single-doped steel fibre and single-doped nanometre fibre and steel-Nano hybrid fibre after data processing, and the graphs were plotted for each of them.

Table 3. Basic parameters of the specimen, cross section (100 ×100 mm), diameter of reinforcing steel (14 mm), and reinforcement rate (1.58 %).

Number	Steel fibre Volume rate (%)	Nanometre fibre Dosage (%)
S0N0	0	0
S05N0	0.5	0
S10N0	1	0
S15N0	1.5	0
S0N05	0	0.05
S0N10	0	0.1
S0N15	0	0.15
S0N20	0	0.2
S05N05	0.5	0.05
S05N10	0.5	0.1
S05N15	0.5	0.15
S05N20	0.5	0.2
S10N05	1	0.05
S10N10	1	0.1
S10N15	1	0.15
S10N20	1	0.2
S15N05	1.5	0.05
S15N10	1.5	0.1
S15N15	1.5	0.15
S15N20	1.5	0.2

3 Analysis of experimental results

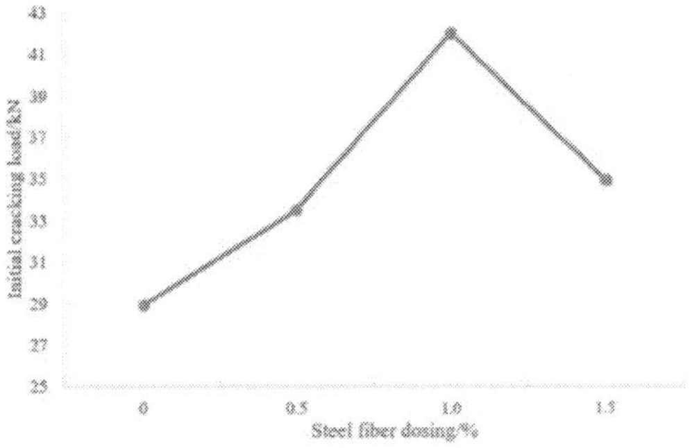

Fig. 1. Initial cracking load of single-doped steel fibre specimens

https://doi.org/10.1051/e3sconf/202234101008

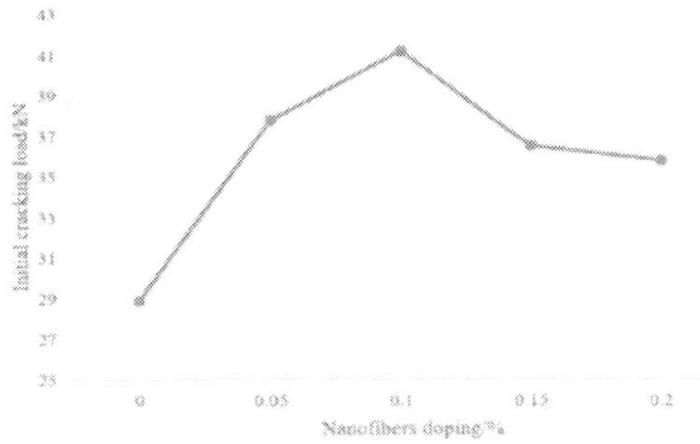

Fig. 2. Initial cracking load of single-doped nanometre fibre specimens

From the above two figures, it can be seen that when steel fibre or nanometre fibre are doped alone, the initial cracking load on the specimens is elevated to some extent.

The initial cracking load increased with the increase of steel fibre dosing for single-doped steel fibre axial tensile specimens at 0%-1%, the initial cracking load showed a decreasing trend when the dosing was increased from 1% to 1.5%. In this test, the initial cracking load reached the maximum at 1% of steel fibre doping, at which time the initial cracking load was increased by 45.3% compared to the plain concrete specimens. The initial cracking load increased with the increase of nanometre fibre admixture at 0%-0.10% for the single-doped nanometre fibre axial tensile specimens, the initial cracking load showed a decreasing trend when the admixture increased from 0.10% to 0.20%. In this test, the maximum initial cracking load was reached at 0.10% nanometre fibre dosing, and at this time, the initial cracking load was increased by 42.6% compared to the plain concrete specimens.

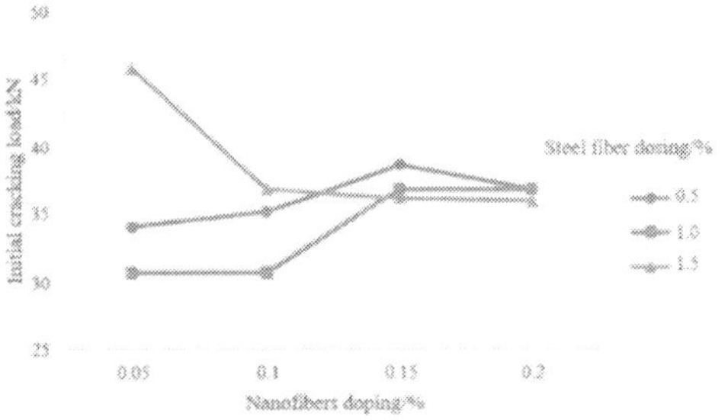

Fig. 3. Relationship between mixed fibre dosing and the initial cracking load of specimens.

As can be seen from the analysis in Figure 3 above, the initial cracking loads of the specimens and the blended fibre admixture do not show a significant regularity. Compared to the case of steel and nano-fibers alone in Figure 1 and Figure 2, the initial cracking load of the steel-Nano hybrid fibre specimens increased but not significantly with the increase of the blended doping of both fibres. The peak was reached at 1.5% of steel fibre and 0.05% of nanometre fibre, at which point the initial cracking load elevation reached 58.9% compared to the plain concrete specimens.

The uneven distribution of hybrid fibers in the concrete matrix leads to defects in the specimen, so the initial crack load may decrease with the increase of the content.Thus, the initial cracking load of the specimens did not show any obvious regularity between the specimens and the number of mixed fibres.

3.1 Effect of steel-Nano hybrid fibre on crack morphology

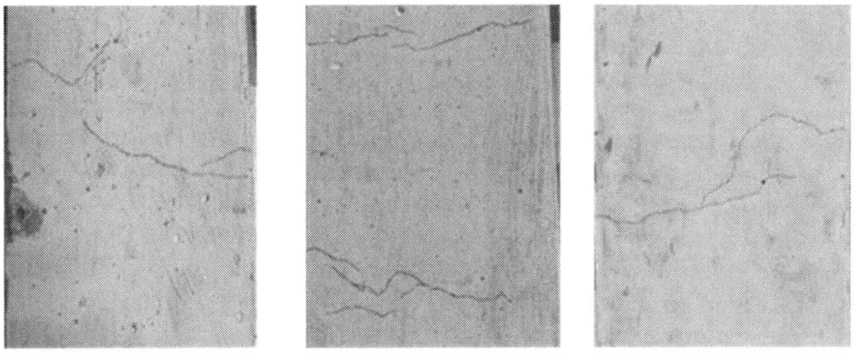

(a) Crack discontinuity. (b) Crack bifurcation. (c) Crack misdirection.

Fig. 4. Local crack pattern diagram

The above figure shows the local crack morphology of the axial tensile specimen, and after observation, it can be found that the crack morphology of the concrete has been significantly changed after the admixture of mixed fibres. The crack development in the specimens is chaotic due to the three-dimensional chaotic distribution of fibres in the concrete, which is blocked by the fibres during cracking. After further observation, it was found that the crack discontinuity, bifurcation, and chaotic direction were more evident in the specimens mixed with hybrid fibre compared to the specimens mixed with steel fibre alone.

To analyze the reason, the incorporated nanometre fibre generates many pin-rod crystal structures between the concrete hydride C-S-H gels, which realize the filling and bridging function of rigid particles. Thus, it can resist deformation and prevent the development of micro fine cracks, making the crack discontinuity, bifurcation, and disorderly direction of the specimen more obvious.

3.2 Effect of steel-Nano hybrid fibre on the average crack spacing

Figure 5 shows the average crack spacing of the axial tensile Specimens corresponding to the nanometre fibre dosing of 0%, 0.05%, 0.1%, 0.15%, and 0.2% of cement mass for several different dosing levels of steel fibre at 0%, 0.5%, 1%, and 1.5%, respectively.

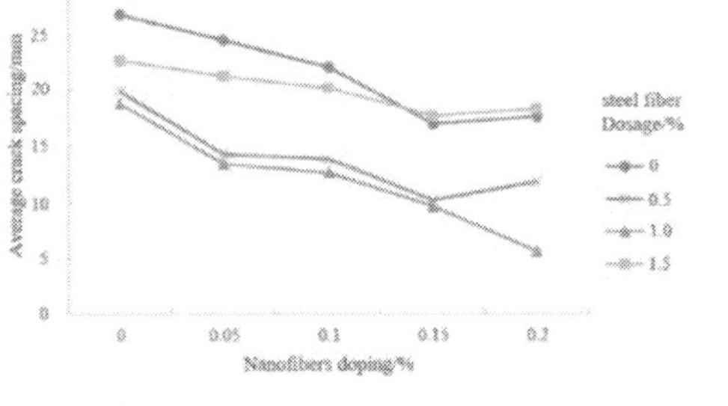

Fig. 5. Average crack spacing of axial tension specimens.

Observing Figure 5, it can be seen that the average crack spacing of the specimens got significantly reduced compared to plain concrete after mixing with mixed fibres. The average crack spacing decreases with the increase of admixture when the steel fibre admixture is 0%-0.15% at a certain amount of steel fibre, however, the average crack spacing appears to increase when the admixture is increased from 0.15% to 0.20%. The average crack spacing decreases with the increase of the dose when the nanometre fibre dose is 0%-1%, however, the average crack spacing increases when the dose increases from 1.0% to 1.5%.

According to the research results [15], the main factor affecting the average crack spacing of reinforced concrete is the ratio of the difference between the tensile strength of concrete and the residual tensile stress on the crack section to the average bond stress between the reinforcement and concrete. Therefore, the reasons for this are mainly the following two aspects: (i) the admixture of mixed fibres certainly improves the tensile strength of concrete, but the steel fibre can assume the role of tensile stress at the crack section, thus increasing the residual tensile stress on the crack section, so that the difference between the tensile strength of fibre concrete and the residual tensile stress on the crack fracture surface does not change significantly compared with plain concrete, (ii) the nanometre fibre incorporated in this test have a diameter of 20 nm- 50nm, which can play a good role in structurally filling the 20nm-150nm micro-pores in the cement paste, reducing the porosity of the material, making the cementations material more and more dense, which in turn improves the average bond between the reinforcement and the concrete, thus reducing the average crack spacing of the specimen. When the fibre dosing is too high, the agglomeration phenomenon is easy to occur, which makes the internal defects of the specimen make the average crack spacing of the specimen increase instead, that is, the negative mixing effect occurs, but the average crack spacing is still smaller than the average crack spacing of the plain concrete specimen without fibre dosing.

3.3 Effect of reinforcement stress on the average crack width

The load loading step was taken as 0.1 times of the load corresponding to the yield stress of the reinforcement, and the crack widths of the specimens under all levels of loading from the beginning to the end of loading were measured, and the average crack widths measured for all specimens were processed by the data and divided into three groups: single-doped steel fibre, single-doped nanometre fibre, and mixed steel-nanometre hybrid fibre, and plotted as the following graphs.

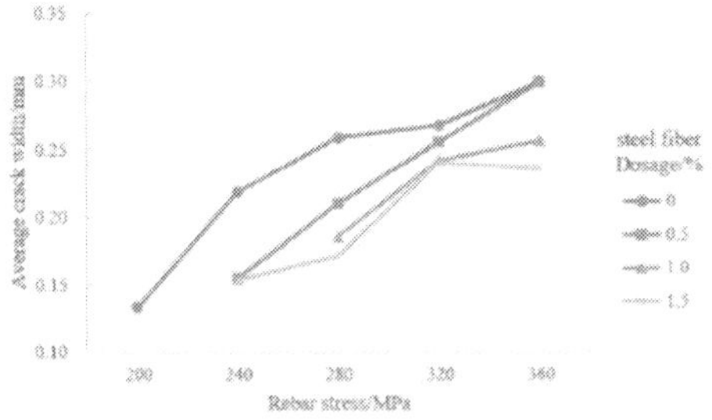

Fig. 6. Steel stresses and average crack width of single-doped steel fibre specimens.

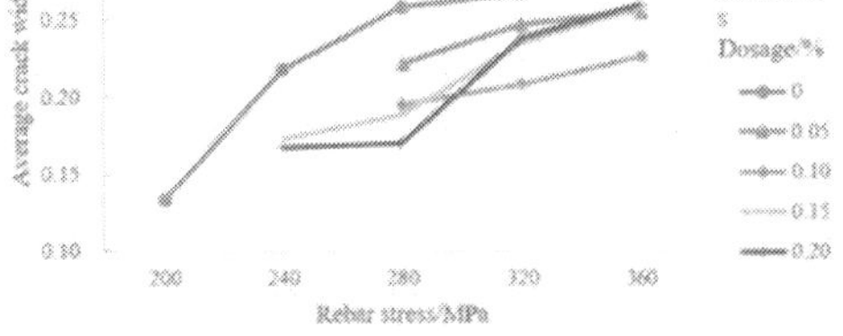

Fig. 7. Stress and average crack width of reinforcement in single-doped nanometre fibre specimens.

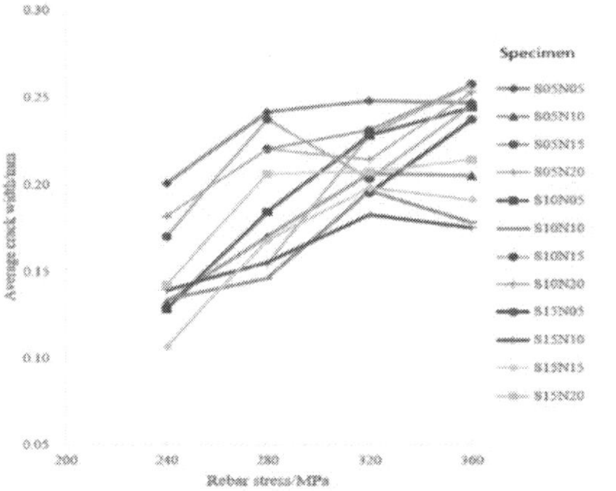

Fig. 8. Mixed fibre specimens reinforcement stress and average crack width,cross section (100 ×100 mm), diameter of reinforcing steel (14 mm), and reinforcement rate (1.58 %)

From the above graphs, it can be seen that the average crack width of the specimens in all three cases of single-doped steel fibre, single-doped nanometre fibre, and mixed steel-Nano hybrid fibre show a linear relationship with the steel stress. It is found in Figure 6 that the average crack width of the specimen is the smallest when the steel fibre admixture is 1.5% for the single-doped steel fibre specimen, i.e., the optimal crack limiting effect, so the optimal admixture of steel fibre is 1.5%, it is found in Figure 8 that the average crack width of the specimen is the smallest when the nanometre fibre admixture is 0.1% for the single-doped nanometre fibre specimen, i.e., the optimal crack limiting effect, so the optimal admixture of nanometre fibre is 0.1%.

To better explore the effect of different blended fibre doping on the crack limiting effect of the specimens, the average crack data at two steel stress levels of 320 MPa and 360 MPa were selected, and the relationship between the blended fibre doping and the average crack width of the specimens at the two steel stress levels were plotted as follows.

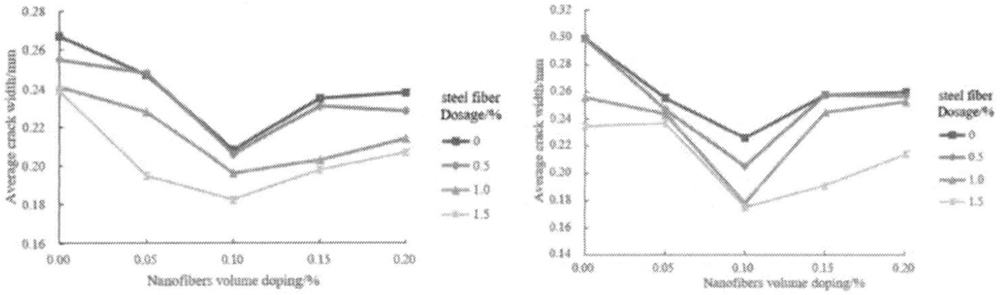

(a) Reinforcement stress of 320 MPa. (b) Reinforcement stress of 360 Mpa.

Fig. 9. Average crack width of specimens under the same reinforcement stress.

Analysis of the above graph shows that at the stress levels of 320 MPa and 360 MPa for reinforcement stress, the average crack width of the specimen decreases with the increase of nanometre fibre doping when the nanometre fibre doping is between 0% and 0.10%, when the nanometre fibre doping is greater than 0.10%, the average crack width of the specimen increases with the increase of nanometre fibre doping at this time. At fixed nanometre fibre dose, the average crack width of the specimen decreased when the steel fibre dose increased from 0% to 1.0% at the level of steel stress 320 MPa, and the average crack width of the specimen increased when the steel fibre dose increased to 1.5%, but it was still smaller than the average crack width of the specimen with single-doped steel fibre, at the level of steel stress 360 MPa, the average crack width of the specimen decreased with the increase of the steel fibre dose. The average crack width of the specimens decreased with the increase of steel fibre doping at the level of reinforcement stress 360 MPa. At the stress levels of 320 MPa and 360 MPa, the steel fibre and nanometre fibre showed good synergistic crack resistance.

The crack control of the specimens with steel-Nano hybrid fibre was also better than that of the specimens with steel fibre alone or with nanometre fibre alone at higher reinforcement stress levels (360 Mpa). Compared with the previous axial tensile tests[18] of steel-polypropylene hybrid fibre, which showed a negative mixing effect at higher steel stress levels, the steel-Nano hybrid fibre showed better crack control at higher steel stress levels, and the most significant crack control effect was achieved at 0.10% nanometre fibre and 1.5% steel fibre.

3 Conclusions

After preliminary test and later data analysis, the following conclusions are obtained:

The addition of steel-Nanometre hybrid fibre can significantly change the crack shape, increase the number of cracks, and effectively prevent the crack penetration.

The addition of steel-Nanometre hybrid fibre significantly increases the initial crack load of the specimen, and the maximum increase is 58.9% higher than that of plain concrete.

The average crack spacing of the specimen decreases obviously with the addition of steel-Nanometre hybrid fibre.

Before the yield of reinforcement, the average crack width of the specimen is basically positively correlated with the reinforcement stress. Under the high reinforcement stress level(360MPa), the crack control effect of the specimen mixed with steel-Nanometre hybrid fibre is also better than that of the specimen mixed with single fibre.

Acknowledgments

The authors gratefully acknowledge the fund support from the National Natural Science Foundation of China (51879146).

References

1. *Code for design of pressure steel pipes for hydropower plants: NB/T35056-2015* (2016).
2. Jiamei Zhou, Qian Zhang, Guowang Meng, Railway Standard Design, **61**, 84-90 (2017).
3. Qingguo Cheng, Yunxian Xu, Zuwen Lu, China Railway Science, 3-11 (1999).
4. A. E. Naaman, S. P. Shah, Journal of the Structural Division, **102**, 1537-1548 (1976).
5. R.N. Swamy, International Journal of Cement Composites, **2**, 43-53 (1980).
6. Chengkui Huang, Guofan Zhao, Building Structure,74-79 (2005).
7. Junming Yang, Xin Cao, Building Materials Development Guide, **19**, 61-62,.(2021).
8. Tianlong Qiao, Sichuan Cement, 35-36 (2021).
9. Ruipin Sun, Building Materials World, **36**, 6-9 (2015).
10. Tao Ji,Yvzhou Huang, Z. Q. Zheng, Concrete, 13-15 (2003).
11. Wei Sun, Hongping Qian, Huisu Chen, Journal of Silicates, **28**, 95-99 (2000).

12. Yi Li, *Hybrid fibre concrete crack resistance and toughening and durability performance* (2012).
13. Biao Huang, Qilin Lu, Journal of Forestry Engineering, **1**, 1-9 (2016).
14. D. Klemm, F. Kramer, S. Moritz, Angewandte Chemie-International Edition, **50**, 5438 (2011).
15. Chengkui Huang, Guofan Zhao, Keshun Lv, Journal of Building Structures, **18**, 1-10 (1997).
16. De menezes A. J., Siqueira G., Curvelo A. A. S., Polymer, **50**, 45-52 (2009).
17. Xuefeng Yang, Hailin Wu, Yanan Liu, Hydropower, **45**, 113-117 (2019).
18. Hailin Wu, Ziqiang Pei, Xuefeng Yang. Journal of Huazhong University of Science and Technology (Natural Science Edition), **48**, 43-47 (2020).

Performance of cement stabilized recycled aggregate base course added with metakaolin

Jianming Yao[1], *Wei* Wang[2], and *Xuyan* Song[3,*]

[1]Suzhou Transport Investment Planning Design and Management Co., Ltd., 215000, China
[2]School of Civil Engineering,Chongqing Jiaotong University,Chongqing 400074,China
[3]Road Engineering Research Center, Suzhou University of Science and Technology, 215000, China

Abstract. In order to better use the old cement concrete pavement materials in the cement stabilized base of the reconstruction or new road. In this paper, metakaolin was used to replace recycled fine aggregate with different proportions to prepare cement stabilized recycled aggregate base mixture, and the physical and mechanical properties and dry shrinkage properties of cement stabilized recycled aggregate base mixture were studied. The results of the study show that adding a proper amount of metakaolin can reduce the optimal water content of the mixture, increases the maximum dry density and 7d unconfined compressive strength of the mixture, and also reduces the total water loss and dry shrinkage of the mixture. The research results can provide new ideas for the recycling of road waste materials.
Keywords: Metakaolin; Recycled aggregate; Strength; Dry shrinkage.

1 Introduction

Since the 40 years of reform and opening up, China has devoted all its manpower and material resources to civil engineering and greatly improved the quantity and quality of infrastructure. The vigorous development of road traffic and transportation projects has contributed to China ' s economic take-off. In the early and middle stages of this construction process, most roads use cement concrete as road surface layer. After years of vehicle load and natural factors, many urban roads have been damaged to varying degrees. The process of road reconstruction or expansion will produce a large number of waste solid waste, disposal of these wastes need to spend a lot of money and manpower, adversely affect the surrounding environment of the city[1].

Cement stabilized base is a kind of base structure widely used in road in China because of its high stiffness, low cost and good construction performance. The thickness of the structural layer of the base material is generally large, which requires more sand and stone materials. Therefore, the application of solid waste as recycled aggregate in cement stabilized base is one of the best ways to realize the reuse of solid waste, which not only saves costs, but also has the effect of protecting the environment [2,3]. However, due to the large amount of cement paste contained in recycled aggregate and the micro-cracks

* Corresponding author: songxuyan2002@163.com

generated in the crushing process, the water absorption rate increased, the mechanical properties decreased, and the cracks were prone to dry shrinkage, which further led to the occurrence of pavement reflection cracks in the service process.

Metakaolin is a mineral admixture made by calcining kaolin. It contains a large amount of amorphous silicon dioxide and aluminum oxide. It has high pozzolanic activity and the function of filling the gap of cement paste, which can improve the mechanical properties of concrete significantly reduce its internal porosity [4-5]. Wang Chong et al. [6] showed that adding 5% metakaolin to concrete can significantly reduce the shrinkage of concrete, and the shrinkage of concrete has a great relationship with the content of metakaolin. In order to explore whether metakaolin can improve the mechanical properties and drying shrinkage performance of cement stabilized recycled aggregate base, metakaolin was added to cement stabilized recycled aggregate base mixture with different contents, and the changes of physical and mechanical properties and drying shrinkage performance were studied.

2 Raw materials and test methods

2.1 Raw material

Aggregate: The recycled aggregate used in this experiment was taken from the pavement recycling material of a highway in Jiangsu. It is divided into recycled coarse material (10~26mm), recycled medium material (5~10mm) and recycled fine material (0~5mm). The apparent densities were 2.62g/cm^3, 2.65g/cm^3 and 2.81g/cm^3, respectively. The water absorption rates were 5.28 %, 8.09 % and 10.1 %, respectively. The crushing value was 24.5, the needle content was 6.0 %, and the mud content was 0.7 %. In order to analyze and study the composition and surface morphology of recycled aggregate, XRD and SEM were used to observe the recycled aggregate. The diffraction pattern and microstructure are shown in Fig. 1 and Fig.2.

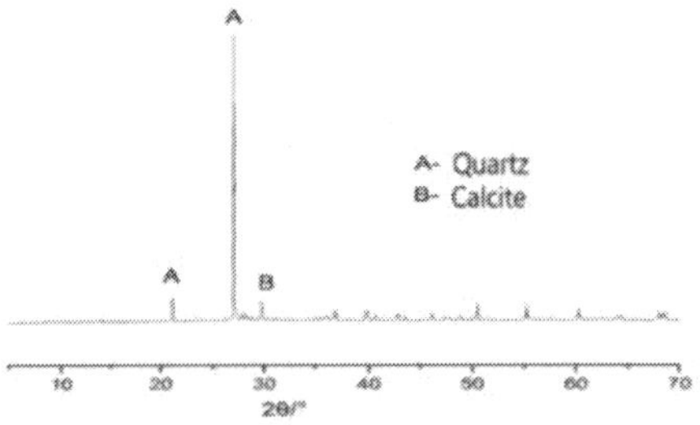

Fig. 1. XRD diffraction pattern of recycled aggregate.

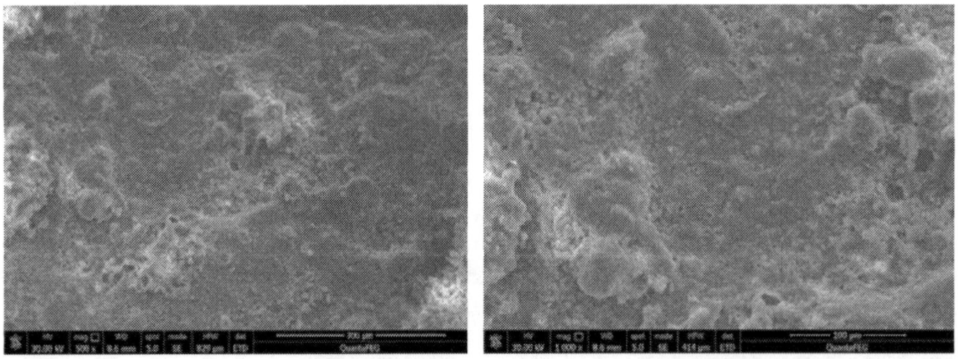

Fig. 2. SEM of recycled aggregate.

It can be seen from the above spectra that the main components are quartz (SiO_2) and calcite ($CaCO_3$), and the quartz content in aggregate is the highest. It can be seen from the SEM diagram that the grain size of recycled aggregate is different, the surface is rough, and a large number of hydration products such as cement mortar are covered. The porous interface proves that recycled aggregate has high water absorption. In addition, obvious microcracks can be observed on its surface, which may adversely affect its strength.

The cementitious material is Tianshan PO42.5 cement produced by Suzhou Damin Building Materials Co., Ltd. Metakaolin is produced by Henan Jiaozuo Yukun Mining Co., Ltd. Chemical composition is shown in Table 1; the mixing water is Suzhou municipal drinking water.

Table 1. Chemical composition of metakaolin.

Sio_2(%)	Al_2o_3(%)	Fe_2o_3(%)	Tio_2(%)	CaO(%)	MgO(%)	K_2O(%)	Na_2O(%)
55.06	44.12	0.76	0.24	0.17	0.06	0.55	0.06

2.2 Test method

The recommended gradation of cement stabilized graded crushed stone in JTG / T F20 - 2015 ' Highway Pavement Construction Technical Rules ' is adopted. Five schemes are designed according to the C-B-1 gradation required by highways and first-class highways : the original base mixture (G0), and the metakaolin is replaced by 1 %, 3 % and 5 % of the total mass of the mixture to replace the fine aggregate in the mixture, respectively represented by G1, G3 and G5. In addition, the metakaolin of 1 % of the total mass of the mixture is used to replace 0.075 mm ultrafine aggregate (G1X) in the mixture, and the cement content is 3.5 %. The compaction test, unconfined compressive strength test and dry shrinkage test are carried out in accordance with the ' Test Specification for Inorganic Bond Stabilized Materials in Highway Engineering (JTG E51-2009) '.

3 Test result analysis

3.1 Physical and mechanical properties

The optimum water content, maximum dry density and 7 d strength of each mixture were obtained through compaction test and unconfined compressive strength test. The test results are shown in Table 2.

Table 2. Physical and mechanical properties of cement stabilized recycled aggregate base mixture with different metakaolin content.

	G0	G1	G3	G5	G1X
Optimum moisture content (%)	11.198	10.480	10.784	11.027	12.640
Maximum dry density	2.094	2.132	2.121	2.103	2.159
7d unconfined compression strength (Mpa)	5.02	5.20	5.35	4.84	5.31

It can be seen from Table 2 that the change of the optimum moisture content of the base mixture is related to the content of metakaolin added. Using metakaolin instead of fine aggregate in base mixture can reduce the optimum moisture content. After adding 1 % metakaolin, the decrease was the largest, and the optimal moisture content decreased from 11.198 % of G0 to 10.499 % of G1, with a decrease of 6.41 %. However, the optimum moisture content of G1X increased by 12.9 % than that of G0 after 1 % metakaolin was replaced by superfine aggregate in the mixture. In addition, after adding a small amount of metakaolin to the base mixture, the maximum dry density of the mixture increased slightly, basically between 2.10 g / cm3 and 2.16 g / cm3, with an increase of 0.5 % – 3.3 %.

It can be seen from Table 2 that the 7d unconfined compressive strength of the base mixture is related to the content of metakaolin added. The 7 d unconfined compressive strength of G1, G3 and G1X were higher than that of G0, with the increase of 3.6 %, 6.6 % and 5.8 %, respectively. When 5 % metakaolin was added to the base mixture, the 7d unconfined compressive strength of G5 decreased by 0.18 MPa, which was 3.4 % lower than that of G0. It is proved that adding a small amount of metakaolin can improve the 7d unconfined compressive strength of the base mixture, which is mainly because the strength of the base mixture is related to its composition and pore structure characteristics. Metakaolin belongs to mineral admixtures and has activity. The main active components are alumina and silica, accounting for about 90 %. The size of active components is small, both below 5 um. Therefore, they have good volcanic ash effect when used as mineral admixtures. Metakaolin can exert pozzolanic effect at the early stage of cement hydration, thereby reducing the content of calcium hydroxide crystal in hydration products and increasing the content of C-S-H gel. At the same time, pozzolanic effect can reduce the calcium-silicon ratio in the gel, thereby improving the density of hydration products. In addition, metakaolin can play a size effect in the mixture, generate solid particles and dense accumulation, so as to refine the pore size distribution of the mixture and improve the compressive strength of the mixture. Due to the existence of a large number of glassy alumina in metakaolin, the volcanic ash effect is more intense in the early stage, which has a good promoting effect on early strength.

3.2 Drying shrinkage

According to the previous research results, the shrinkage performance of cement stabilized base mixture has a certain relationship with the optimum moisture content. According to the compaction test results, G1, which is the most economical and has the best effect on reducing the optimum moisture content, is selected for the dry shrinkage test in this section, and the results are compared with those of G0. The test results are shown in Figs. 3.

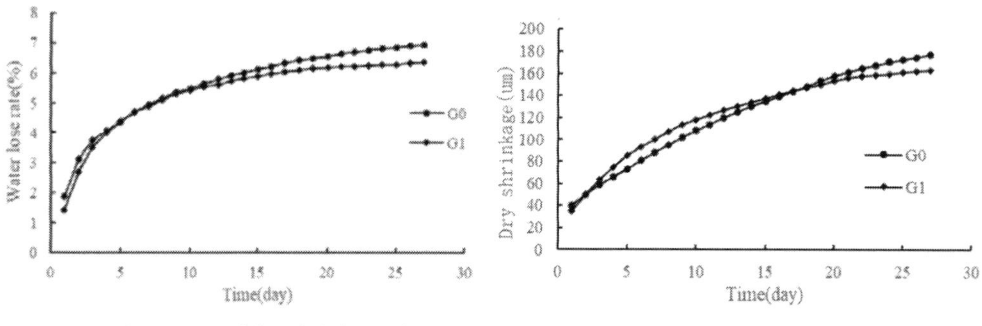

Fig. 3. Water loss rate and dry shrinkage change over time.

It can be seen from Fig. 3 that the water loss rates of the two base mixtures increase with time, and the water loss rate is faster in 1 – 7 days, and slightly slower in 7 – 14 days. In the late stage of water loss, namely after 14 days, the water loss rate remained basically stable. After adding 1 % metakaolin, the variation trend of water loss rate of G1 with time was almost consistent with that of G0. However, the 27 d total water loss rate of G1 was 6.36 %, which was 8.6 % lower than that of G0.

The change trend of dry shrinkage of the two kinds of base mixtures with time is basically the same. The dry shrinkage increases with time, and the growth is fast in the early stage, and gradually decreases in the late stage. After adding 1 % metakaolin, the dry shrinkage of G1 was greater than that of G0 in the first 14 days, but the total dry shrinkage of 27 days was less than that of G0. The dry weight of G1 was reduced by 14um than that of G0, and the reduction ratio was 7.9 %. This is because the dry shrinkage of the base mixture is usually caused by the water loss of the capillary pores and gel pores smaller than 50 nm. Metakaolin can reduce the content of capillary pores in the mixture and increase the content of gel pores. The role of metakaolin in refining the pore size can reduce the total dry shrinkage of the base mixture. In addition, metakaolin can also promote the hydration process of cement, thereby increasing the drying shrinkage caused by hydration reaction, which has a negative impact on reducing the early drying shrinkage of base mixture. Therefore, the influence of metakaolin on dry shrinkage of base mixture is caused by the combined action of the above two parts. Metakaolin has large particle size and contains a large number of active substances such as alumina, which can promote the formation of AFt and AFm in volcanic ash reaction and further inhibit the shrinkage of base mixture.

4 Summary

(1) Adding a small amount of metakaolin can reduce the optimum water content of cement stabilized recycled aggregate base mixture and improve the maximum dry density of the mixture.

(2) The unconfined compressive strength of cement stabilized recycled aggregate base mixture can be significantly improved when the dosage of metakaolin is 1 % ~ 3 %, but it is unfavorable to the strength growth when the dosage is increased to 5 %.

(3) The water loss rate and dry shrinkage of cement stabilized recycled aggregate base mixture with metakaolin increase with time, which are higher than those without metakaolin in the early stage, but lower in the later stage, and the total dry shrinkage is smaller. Adding 1 % kaolin can reduce the dry shrinkage of cement stabilized recycled aggregate base mixture.

References

1. Lingqing Liu. Performance and design of cement stabilized recycled aggregate[D]. Chang'an University(2014)
2. Yiwu Shen. Key technology of dry shrinkage performance of recycled concrete road base materials [D]. Suzhou University of Science and Technology(2020)
3. Zenghua Zhou.Feasibility analysis of construction waste recycling in road construction[J]. *Engineering and Construction*, **23**,624-626(2009)
4. Tianxiang Cui. Experimental Study on Performance of High Performance Concrete with Metakaolin[D]. Central South University of Forestry and Technology(2021)
5. Qiang Yu,Junjie Zeng,Zhihong Fan. Comparative analysis of the influence of metakaolin and silica fume on concrete performance[J]. *Bulletin of The Chinese Ceramic Society*, **33**,3135-3139(2014)
6. Chong Wang,Xianwen Wu,He Huang. Performance comparison of high strength concrete prepared by metakaolinite and silica fume[J]. *Journal of Chongqing University*, **30**,65-69(2007)

Research on hydrogen fuel application (in aircraft) based on carbon neutralization

Yan Zhang[1,*], and *Liang* Sun[2]

[1]Economy and Management Department, Civil Aviation Management Institute of China, Beijing 100102, China
[2]Office of Academic Research, Civil Aviation Management Institute of China, Beijing 100102, China

Abstract. Hydrogen fuel has the advantages of green environmental protection and high energy density, hydrogen aviation is also developing rapidly in China. Due to the high-quality characteristics of hydrogen, hydrogen aviation is the key to zero pollutant emission and sustainable development of aviation industry in the future. In the context of "carbon neutralization", the EU and other countries have launched the promotion of large-scale green hydrogen plan focusing on the aviation industry. Although they are still facing some technical bottlenecks to break through some problems of hydrogen fuel in their development, hydrogen fuel will be an important means to achieve low-carbon and environmental protection in the post epidemic era.
Keywords: Hydrogen fuel; Hydrogen powered aircraft; Carbon neutralization; Challenges and suggestions.

1 Introduction

Under the background of realizing "carbon neutralization" to deal with global climate change, the global carbon neutralization action is accelerating the tide of green and low-carbon transformation of the world environmental protection. Sweden's carbon peak in 1970 was 92.29 million tons, and the United States' carbon peak in 2005 reached 6130.55 million tons. By 2019, 46 countries and regions in the world had reached the peak, mainly developed countries.

There are three possible means to reduce emissions in the aviation field: improving energy efficiency, carbon capture technology and the use of alternative energy. According to the prediction of ICAO, simply improving energy efficiency can not achieve the goal of carbon neutralization, and finding alternative energy is an inevitable choice in the long term. Due to the high energy supply requirements of aircraft flight and the inability to charge midway, the battery at this stage can not solve this problem, and hydrogen fuel has high calorific value, fast hydrogenation speed and high mass power density, or becomes the key to solve this problem. At present, the application of hydrogen energy in aviation is still in the experimental stage all over the world. Although the United States, Europe, Russia and other countries have carried out relevant research and flight test, wide range application is

[*] Corresponding author: zhangyan1027108@126.com

still a long-term goal. As the world's largest aircraft manufacturer, Airbus takes hydrogen fuel as the most promising energy type and renewable energy to provide power for aircraft and aviation. In 2020, it announced three hydrogen energy concept aircraft, becoming the world's first zero emission aircraft, code named zeroe, which will change the future air transportation mode and be launched in 2035.

Under the background that China has achieved the goal of "carbon neutralization" in 2060, hydrogen energy as a secondary energy is indispensable. Due to the limited options for emission reduction in the field of civil aviation, hydrogen fuel as a way to reduce carbon emissions will become the key strategy and long-term key path of the net zero goal. In 2019, the North research center of COMAC and the hydrogen energy technology development company of State Power Investment Group launched the development of "lingque H" new energy verification machine, which was successfully tested at Zhengzhou Shangjie airport. In March 2019, Dalian Institute of chemistry, Chinese Academy of Sciences and Liaoning General Aviation Research Institute jointly developed the first manned hydrogen fuel cell test machine, which successfully flew in Shenyang [1-2].

After exploration in the past few years, all countries urgently hope to accelerate the application of hydrogen energy in aviation, especially in the wide range field of aviation.

2 Characteristics of hydrogen fuel

Hydrogen, with the element symbol H, is the lightest chemical element. It is a gas at normal temperature and pressure. As the most common element, it is mainly stored in water. Since NASA technicians successfully used hydrogen as fuel for the Saturn V lunar rocket, fuel experts are more convinced that hydrogen can be used as an ideal fuel to replace coal, natural gas and oil, and believe that mass production of hydrogen can be achieved, even in atomic power stations. Hydrogen is considered as one of the major energy solutions of the twenty-first century, capable of meeting future energy needs. Hydrogen is clean, sustainable fuel and it has promise as a future energy carrier. It also has the ability to substitute the present energy infrastructure which is based on fossil fuel. Being 61a zero-emission fuel, it could reduce environmental impacts and craft novel energy opportunities. [3-4]

Among the renewable sources that are considered to produce hydrogen, wind and solar energies are most commonly used for hydrogen production up to date. Most of the methods for the production of green hydrogen consider obtaining hydrogen either through water splitting reactions (e.g. water electrolysis, water thermolysis, photocatalytic water splitting, and thermochemical water splitting) or via dehydrogenation of hydrogen carrier molecules. [5]The chemical energy of hydrogen can be converted into thermal energy or electric energy, so one of the wide applications of hydrogen is as aerospace fuel. The different characteristics of hydrogen determine that hydrogen has different application technologies [6].

2.1 Buoyancy of hydrogen

As hydrogen is the lightest of all gases, its density is only 0.0899kg/m^3 at 0 °C and its density is small, so hydrogen has the largest buoyancy and can be used as lifting power.

2.2 High energy properties of liquid hydrogen

In any chemical fuel, the specific energy (unit mass energy) of hydrogen is the largest, and the low calorific value is as high as 120mj / kg, which is about 3 times that of gasoline and kerosene and 4.5 times that of carbon. The power and range of aeroengine are directly proportional to the calorific value of fuel. Using liquid hydrogen as aviation fuel can greatly improve all performance parameters of aircraft, and the takeoff weight of supersonic aircraft fueled with liquid hydrogen is only half of that of kerosene. Lockheed has conducted comparative experiments on aviation kerosene and liquid hydrogen. It is found that under the same payload and range, liquid hydrogen fuel is much lighter, the total weight of the aircraft can be reduced, and the running time can be shortened, thus saving the total fuel consumption.

2.3 Environmental emission characteristics of hydrogen

As a carbon free fuel, hydrogen does not produce CO, CO_2 and CH compounds emitted during hydrocarbon combustion, and will not produce greenhouse effect. Hydrogen is a colorless, odorless, odorless and non-toxic clean fuel. Burning in pure oxygen only produces water, which returns to nature and maintains ecological balance. Advanced hydrogen fuel cell and catalytic hydrogen technology only discharge water. Using these hydrogen energy to drive vehicles and aircraft can achieve real "zero emission". The United States has held the vehicle competition with the least air pollution, and it is concluded that the hydrogen engine won the first place, and the exhaust gas is even cleaner than the inhaled urban air.

2.4 Hydrogen fuel resources

As an important characteristic of new energy, it is rich in resources and wide sources. The earth's atmosphere contains 1.6 billion tons of hydrogen, but free hydrogen is difficult to use. At present, more than 90% of the world's hydrogen fuel is produced by fossil fuels, and part of it is produced by electrolysis. With the development and breakthrough of hydrogen energy technology, electrolytic water and dissociated steam will be gradually transformed, and the storage capacity of water in Shanghai is 1.3×10^{18} tons, water can be recycled. Therefore, hydrogen energy is not only inexhaustible, but also inexhaustible. According to the investigation, there are 1 billion tons of water on the moon, and Jupiter is 318 times that of the earth. It is a large natural warehouse of hydrogen resources, which will be developed and utilized by mankind one day.

3 Challenges of hydrogen fuel application in Civil Aviation

Since the invention of aircraft in the early 20th century, mankind has not stopped the exploration and application of new technologies. However, hydrogen energy has attracted more and more attention because of its superior properties. Therefore, although the application of hydrogen energy to aircraft has a series of obvious advantages in theory, there are still many problems to be solved if it is really put into wide range application.

3.1 Production, storage and use of hydrogen fuel

Compared with lithium battery pack as battery system, hydrogen fuel can prolong the service life by 10 times and increase the energy density by 5-20 times. Hydrogen energy has unique application prospects. However, over the past few decades, the aviation industry has had a comprehensive understanding of aviation kerosene used by turbine and turbofan engines and accumulated rich experience in safe storage, transportation and use of these system fuels. The production, storage and safe and effective use of hydrogen fuel need the joint cooperation of government regulators and manufacturers to achieve innovative breakthroughs in management measures and key technologies. At present, there are still many unknown fields in the related theory, knowledge and application of hydrogen energy [7].

3.2 Infrastructure and supporting facilities

Supporting the propulsion system of hydrogen powered aircraft requires new ground processes, such as filling hydrogen fuel in the shortest possible time to make the parking time of the aircraft as short as possible. How to ensure the safety of these processes.
The operation of hydrogen powered aircraft must be determined at an early stage and take into account the construction of infrastructure. At present, both airlines and airport supporting facilities are prepared for aviation kerosene. Therefore, the main difficulties in promoting hydrogen energy include the lack of infrastructure.

3.3 Policy support

To achieve the goal of "carbon neutralization" and zero carbon dioxide emission as soon as possible, we need not only the support of technology research and development, but also the support of policies.

In the United States, 85% of flights are concentrated in 40 airports, and the remaining 15% are distributed in more than 450 airports. The larger the energy filling network, the more difficult it is to replace it as a whole, and the higher the cost. Each smaller station needs a more complex fuel logistics support system, and each kilogram of fuel allocated will produce a higher amortization cost. At the same time, The overall upgrading of the fuel system requires a lot of capital and human investment. It is difficult to imagine without the support of local governments.

4 Enlightenment and suggestions of hydrogen aircraft

At present, according to the research on hydrogen powered aircraft by major research institutions and enterprises and the judgment of relevant development trends, the following enlightenment is given to the technological development of aviation "carbon neutralization" in China as soon as possible [8].

4.1 Develop air route map

Facing the challenge and urgency of decarbonization in the aviation industry, it is very necessary to formulate a systematic strategy and a clear aviation roadmap. It is difficult to put hydrogen powered aircraft into use in the short term and faces great challenges. From the perspective of China's current development, the aviation industry is very key and promising in realizing carbon neutralization. Vigorously developing hydrogen aviation technology is not only very conducive to achieving emission reduction targets and promoting the early realization of carbon neutralization, but also driving the systematic innovation and improvement of aviation technology. In order to achieve the goal of Green Aviation in China as soon as possible, it is suggested to formulate the corresponding road map as soon as possible and speed up the performance of relevant technologies.

4.2 Accelerate the exploration and research of hydrogen powered aviation technology

The application potential of hydrogen propulsion technology is encouraging. Research and innovation are very important to realize the potential of hydrogen technology in aviation decarbonization. Europe, America and other countries have carried out various research on hydrogen aircraft technology. China has also carried out relevant research and is in a leading position in some fields. In order to continue to play our role, it is recommended that government departments strengthen organization, systematically plan and study the design technology, fuel supply chain technology and infrastructure development of hydrogen energy to promote aircraft, scientifically plan the application of hydrogen energy in aviation industry and break through various key technologies.

4.3 Speed up the exploration and research of new aircraft aerodynamic layout

In order to give full play to the advantages of new energy propulsion technology, innovative aerodynamic layout research is carried out for the effective integration of airframe and propulsion system to improve the overall efficiency of the aircraft. At present, there is still a certain gap between the relevant research carried out in China and the engineering application in terms of breadth, depth and systematicness. The government departments should strengthen the organization, and the research institutions can carry out systematic planning research on the introduction technologies such as new start-up layout, and pay attention to the sustainability and inheritance of the research results.

5 Conclusion

Although the production cost of hydrogen is high, there is no complete hydrogen transmission pipe network, high transportation cost and safety, there are still many difficulties to be overcome, but the development prospect of hydrogen fuel cell is good. With the decarbonization of aviation industry and the support of various countries for the development of hydrogen energy, the research of hydrogen fuel in aircraft in aviation industry is worth looking forward to. Through the planning and formulation of aviation roadmap, accelerating the exploration of hydrogen powered aviation technology, the exploration of new aircraft aerodynamic layout and aircraft life cycle assessment research, hydrogen fuel will become an important path to achieve low-carbon and environmental protection in the future.

References

1. Min Sun, Large aircraft **9**, 45 (2020)
2. Yangjun Zhang, etal, Aerospace Power **1**,20(2021)
3. Manish Kumar Singla, etal, Environmental Science and Pollution Research **28**, 1560(2021)
4. Shashi Sharma, etal, Energies **14**, 7389(2021)
5. SelmaAtilhan, etal, Current Opinion in Chemical Engineering **31**, 100668(2021)
6. Zhengju Chen, Deping Wang, Journal of Shenyang Institute of Aeronautical Engineering **27**, 88 (1994)
7. Guoxiong Dong, Large aircraft **10,** 24 (2020)
8. Miaoxiang Wang, Aerospace Power **1**, 16 (2021)

Experimental study of liquid droplets impact on powder surface: The application of effective dimensionless parameters in analysis

Siqi Li*, *Hourong* Yu, and *Haisheng* Fang

Huazhong University of Science and Technology, School of Energy and Power Engineering, 1037 Luoyu Road, Wuhan, China

Abstract. Spreading dynamics of liquid droplets impacting onto powder bed are experimentally studied using high-speed photography. Dimensionless numbers—We, Re, the modified We^* and Re^* corrected by substrate deformation—are used to analyze the impact behaviors of droplets. The spreading time and the maximum spreading factor are further analyzed. The spreading time is accurately described by a universal scaling law that is obtained from the modified dimensionless time vs. the effective Weber number (We^*), and the maximum spreading factor is found to follow the modified classic scaling law $\beta_{\max} = f(We^*, Re^*)$.
Keywords: Maximum spreading, Effective Weber number, Spreading time, Droplet impact.

1 Introduction

Droplet impact on powder beds occurs extensively in industrial and technological applications, such as granulation in the pharmaceutical industry[1,2]; granule formation by spraying for powder coating and spray drying[3,4]; and in three-dimensional printing processes[5–7]. It is important to understand the dynamics of the droplet impact process on powder beds, which is critical for controlling such processes. Various impact behaviors—e.g., spreading, retraction, rebounding, and splashing—and the resulting crater morphology as well as, liquid marbles have been investigated in detail[8–10] . The behaviors of droplets upon impact have been studied comprehensively by experimental[11,12], theoretical[13,14], and numerical methods[15].

In the experimental and theoretical study, the impact dynamics of liquid droplets and powder surfaces have been systematically investigated. The influences of initial impact velocity and liquid properties on spreading time and maximum spreading have been examined. By using modified Weber number and Reynolds number, we apply empirical expressions for non-deformable solid substrate to powder bed, and reveal their similarities for maximum spreading and spreading time. Scaling analyses were adopted to reveal three characteristic parameters.

* Corresponding author: m201971069@hust.edu.cn

2 Experimental setup and mathematical models

Alumina beads were supplied by the manufacturer Hebei Xinda Alloy material, China. The initial packing fraction, Φ, was reproducible for successive repeat experiments. And in the experiments, Φ is controlled in the range of 0.58 - 0.60 by using tight compression.

Liquids and their properties used in the experiments are displayed in **Table** 1. Dimensionless parameters used in the analysis are Weber number We, and the Reynolds number Re: $We = \rho D u_i^2/\sigma$; $Re = \rho u_i D/\mu$; where ρ, μ, D, and σ are, respectively, liquid density, liquid dynamic viscosity, droplet impact diameter, and liquid surface tension factor; g is the gravitational acceleration and $u_i = \sqrt{2g(h_r - D)}$ is the velocity at the time of impact. The droplet is released onto powder bed from different heights ranging from 3 to 600 mm, resulting in impact velocity range from 0.2 to 3.2 m/s. Droplets can be considered as sphere upon impact. The equilibrium contact angle, θ, is also measured.

The experimental setup (**Fig. 1**) includes a high-speed video camera (Qianyanlang 5KF20), which can capture the impact, spreading, and penetration in a single video sequence.

Table 1. Physical properties of liquids used.

Liquid	D (mm)	ρ (kg·m^{-3})	μ (mPa·s)	σ (N·m^{-1})	θ(°)
Water	2.10	997	1.003	0.072	86.60
50%ethanol	1.78	900	1.10	0.0279	61.66
Ethanol	1.89	789	1.19	0.0232	66.9
50%glycerinum	2.27	1120	4.21	0.070	57.62
Glycerinum	2.26	1260	1410	0.063	37.70
20% glycol	2.42	1029	1.65	0.0605	66.00
50% glycol	2.41	1073	3.94	0.056	63.87
80% glycol	2.18	1110	10.05	0.0504	68.31
Glycol	2.17	1113	21.00	0.0484	58.23
Acetone	1.66	791	0.306	0.0237	72.77

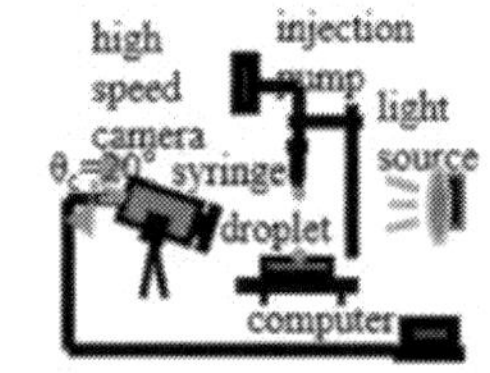

Fig. 1. Schematic of the experimental setup used.

3 Results and discussion

3.1 Spreading time t_s

Fig. 2(a) depicts the spreading time t_s as a function of the impact velocity. The spreading time is typically a few milliseconds, which is of the same order of magnitude as the characteristic advective time[16] $\tau_a = d_0/(2u_i)$ or the capillary-inertial time[17] $\tau_i = (\rho d_0^3/8\sigma)^{0.5}$. **Fig. 2**(b) shows that t_s decreases sharply as the impact velocity increases for $We^* \leq 40$ but slowly decreases for $We^* > 40$. Since τ_a and τ_i ignore influences such as viscosity and surface wettability, a modified capillary-inertia time $\tau_{i'} = (\rho d_{max}^3/8\sigma)^{0.5}$ is adopted to study the spreading[18], where d_{max} is related to liquid viscosity and wettability. **Fig. 2**(c) shows a log-log plot of normalized spreading time $t_s/\tau_{i'}$ vs. We^*. This shows that the function $t_s/\tau_{i'} = \zeta We^{*\lambda}$ gives a good fit to the experimental data. The fitting coefficient ζ and the exponent λ for different viscosities are shown in **Fig. 2**(d). The quantities ζ and λ appear to be independent of liquid viscosity and substrate wettability. Consequently, a universal empirical expression, $t_s/(\rho d_{max}^3/8\sigma)^{0.5} = 1.46 We^{*-0.46}$, can accurately describe the spreading time in powder bed experiments.

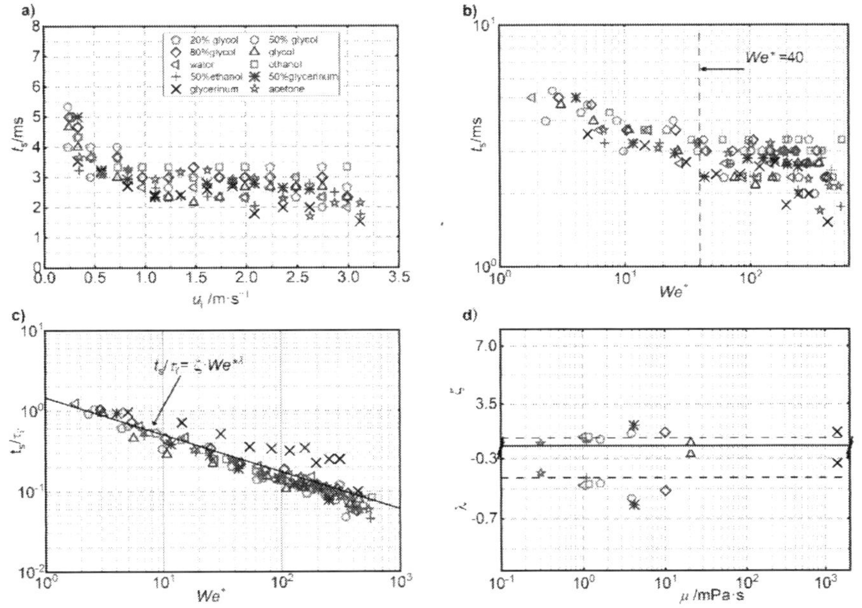

Fig. 2. (a) Spreading time $t_s$vs. impact speed u_i. (b) Spreading time t_s vs. effective Weber number We^*. (c) Normalized spreading time t_s/τ_i, vs. We^*. (d) The coefficient ζ and exponent λ of the graph in part (c) vs. liquid viscosity.

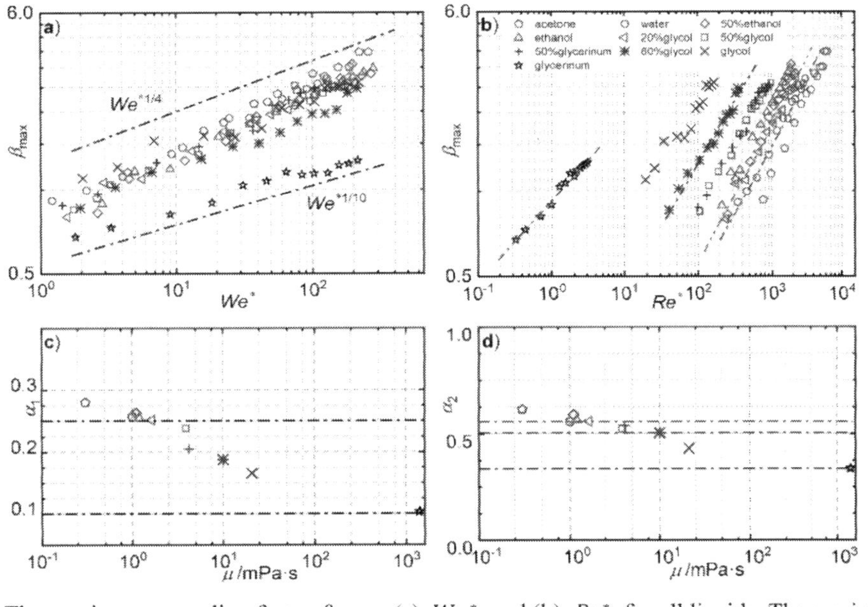

Fig. 3. The maximum spreading factor β_{max} vs (a) We^* and (b) Re^* for all liquids. The maximum spreading factors were fitted in (a) with $\beta_{max} = A_1 We^{*\alpha 1}$and in (b) with $\beta_{max} = A_2 Re^{*\alpha 2}$. α_1 and α_2 as functions of liquid viscosity are plotted in (c) and (d).

3.2 Maximum spreading factor β_{max}

An effective Weber number $We^* = [D/(D + 2Z_m)]We$, where Z_m is the maximum vertical deformation measured[11], is used to replace *We*. It has been shown that We^* collapses the maximum spreading data for various packing densities[19], indicating that We^* can also characterize the effect between liquid droplets with various viscosity and the powder bed with the same packing fraction. **Fig. 3** gives the correlation of β_{max} vs We^* ($We^* = [D/(D + 2Z_m)]We = We^I$) and the correlation of β_{max} vs Re^* ($Re^* = [D/(D + 2Z_m)]Re \neq Re^I$, because the value of $\frac{Re}{Re^I} = \frac{\rho v_p}{\mu} = \frac{2(1-\phi)^4 d_g^2\left[\frac{2\sqrt{6}(D+2Z_m)\sigma^{0.5}\rho^{1.5}u_i D^{-1.5}}{\sqrt{\pi}} + \frac{4\rho\sigma\cos\theta}{d}\right]}{[\mu^2(180\phi^2)]}$ depends heavily on viscosity, Re^I is not applicable to the analysis of viscosity effect in this paper, **Fig. 4** also proves this point) respectively for various viscosity liquids. As shown in **Fig. 3** (a) and (c), a scaling law behavior, $\beta_{max} \sim We^{*\alpha 1}$, is observed at $We^* > 1$. It comes as no surprise to see that β_{max} increases with We^*. The data set exhibits three other distinguished features: (1) The maximum spreading factor is suppressed with increasing viscosities (different color symbols in **Fig. 3** (a)). (2) In the doubly logarithmic scale, there are two asymptotes in the extension of data: the upper is $We^{*1/4}$ and the lower is $We^{*1/10}$. (3) The value of α_1 decreases with increasing viscosities (**Fig. 3**(c)). On the other hand, it seems that the spreading is dominated by viscous force within the liquid for all experimental data. In **Fig. 3**(b), β_{max} grows with Re^* following the power law, $\beta_{max} \sim Re^{*\alpha 2}$.The exponent α_2 is ~0.5 for lowly viscous liquids with $\mu \leq 10$ mPa·s, and it decreases to ~0.33 for liquids with $\mu = 1430$ mPa·s.

Because of the quantitative relation $We^{1/10} \sim D^{-1/4}Re^{1/5}$, a law of the type $\beta_{max} \sim We^{*1/10}$ leads to $\beta_{max} \sim Re^{1/5}$. According to the analysis of the transition to a viscous regime, we thus define the effective number $P^* \equiv We^*/Re^{*4/5}$ for deformable substrate ($P \equiv We/Re^{4/5}$ for non-deformable substrate[20]). The transition between the capillary regime and viscous regime is plotted in **Fig. 4**(a), where the dimensionless viscous extension $\beta_{max}/Re^{*1/5}$ is plotted as a function of P^*. While for $P^\dagger = We^\dagger/Re^{\dagger 4/5}$ in **Fig. 4**(b), the result is surprisingly that the transition is opposite to **Fig. 4**(a)—the viscous droplet spreading changes from viscous regime to the capillary regime as $P^\dagger$ increases. Consequently, $Re^\dagger$—incorporates the bulk wettability of a granular substrate—doesn't fit to interpret the influence of liquid viscosity on spreading. The use of We^* and Re^* reveals the hidden similarities between a droplet impact on deformable substrate and that on a non-deformable substrate for maximum droplet spreading. Same as the non-deformable substrate, the influence of internal inertial flow at maximum spreading phase also exists, and causes the maximum spreading smaller than the theoretical value.

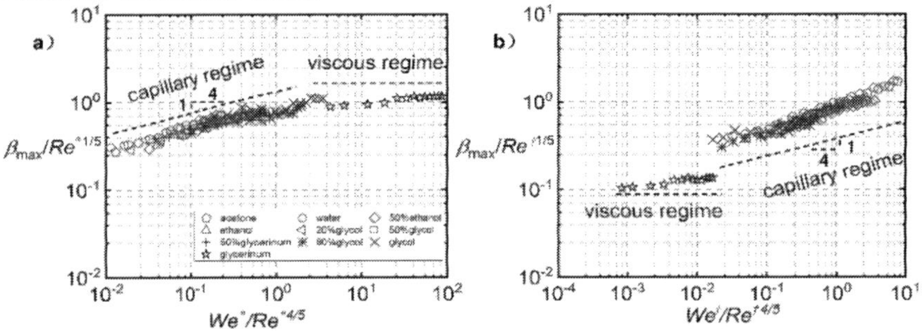

Fig. 4. (a) The maximum spreading factor β_{max} of **Fig. 3** in a doubly logarithmic plot. (b) is based on (a), but $We^*$and Re^* are replaced by $We^\dagger$ and $Re^\dagger$.

4 Conclusions

We have performed an experimental study of various liquid droplets impacting alumina powder beds. Weber numbers (We) and Reynolds numbers (Re) were obtained by adjusted the droplet fall height. Ten liquids were used that provided various viscosities and various surface tensions.

Then we adopted spreading time t_s and maximum spreading factor β_{max} to quantitatively analyze the spreading behavior of the droplets after impacting the powder bed. The modified capillary-inertial time $\tau_{i'} = (\rho d_{\max}^3/8\sigma)^{0.5}$ considering the influence of substrate wettability and liquid viscosity, and the effective Weber number $We^* = [D/(D+2Z_{\mathrm{m}})]We$ considering the influence of the deformable substrate, are innovatively put forward to achieve a universal scaling law: $t_s/(\rho d_{\max}^3/8\sigma)^{0.5} = 1.46We^{*-0.46}$. Besides, the effective weber number We^*, replacing traditional weber number, is adopted to attain the scaling law $\beta_{max} \sim We^{*\alpha 1}$, and α_1 decreases with the increasing viscosity. The same thing happens for the scaling law $\beta_{max} \sim Re^{*\alpha 2}$. The capillary regime considered implies a small P^* ($P^* = We^*/Re^{*4/5} < 1$), the viscous regime refers to large P^* ($P^* = We^*/Re^{*4/5} > 10$). Thus, the use of We^* and Re^* reveals the hidden similarities between a droplet impact on deformable substrate and that on a non-deformable substrate for maximum droplet spreading.

This study is supported by Grants from National Natural Science Foundation of China (No. 51876071).

References

1. Hapgood, K. P. & Khanmohammadi, B. Powder Technol. **189**, 253–262 (2009).
2. Knight, P. C. Powder Technol. **119**, 14–25 (2001).
3. Kayrak-Talay, H. N. E. and D. J. D. L. IFAC Proc. Vol. (2012) doi:10.1002/aic.
4. Moreira, A. L. N., Moita, A. S. & Panão, M. R. Prog. Energy Combust. Sci. **36**, 554–580 (2010).
5. Sachs, E., Cima, M., Williams, P., Brancazio, D. & Cornie, J. J. Manuf. Sci. Eng. Trans. ASME **114**, 481–488 (1992).
6. van Dam, D. B. & Le Clerc, C. Phys. Fluids **16**, 3403–3414 (2004).
7. Van Der Bos, A. *et al.* Phys. Rev. Appl. **1**, 1–9 (2014).
8. Rioboo, R., Marengo, M. & Tropea, C. Exp. Fluids **33**, 112–124 (2002).
9. van der Meer, D. Annu. Rev. Fluid Mech. **49**, 463–484 (2017).
10. Delon, G., Terwagne, D., Dorbolo, S., Vandewalle, N. & Caps, H. Phys. Rev. E - Stat. Nonlinear, Soft Matter Phys. **84**, 3–7 (2011).
11. De Jong, R., Zhao, S. C., Garcia-Gonzalez, D., Verduijn, G. & Van Der Meer, D. Soft Matter **17**, 120–125 (2021).
12. Hardt, S. & McHale, G. Annu. Rev. Fluid Mech. **54**, 83–104 (2022).
13. Li, S. *et al. Science Advances* vol. 7 (2021).
14. Graeber, G. *et al.* Nat. Commun. **12**, 1–7 (2021).
15. Yarin, A. L. Annu. Rev. Fluid Mech. **38**, 159–192 (2006).
16. Supakar, T., Moradiafrapoli, M., Christopher, G. F. & Marston, J. O. J. Colloid Interface Sci. **468**, 10–20 (2016).
17. Wildeman, S., Visser, C. W., Sun, C. & Lohse, D. J. Fluid Mech. **805**, (2016).
18. Lin, S. *et al.* J. Colloid Interface Sci. **516**, 86–97 (2018).
19. Zhao, S. C., De Jong, R. & Van Der Meer, D. Soft Matter **11**, 6562–6568 (2015).
20. Clarke, A., Blake, T. D., Carruthers, K. & Woodward, A. Langmuir **18**, 2980–2984 (2002).

Effect of humification during hydrothermal treatment of Aquatic Wastes with inorganic acid

Huijun Zhang[1,2,*], *Nana* Xie[1], *Zihong* Fan[1], *Tengtun* Xu[1], and *Chunhong* Tang[1,2]

[1]College of Environmental Resources, Chongqing Technology and Business University, Chongqing 400067, China
[2]Chongqing Engineering Research Center for Processing, Storage and Transportation of Characterized Agro-Products, Chongqing 400067, China

Abstract. Taking aquatic processing waste as the main experimental object, to explore the effect of different treatment conditions on the humification process in the acid-hydrothermal reaction process, through the analysis of the hydrothermal process of pH value, organic matter and humus content, humification rate, and infrared spectrum analysis verification to speculate the formation mechanism of humus. The results show that hydrothermal treatment can promote the formation of humus to a certain extent. Under the working conditions of 195°C and 80min, the humus content of aquatic wastes reaches 78.1% and the humification rate reaches 0.35. In addition, within a certain treatment intensity, the hydrothermal humification intensity will increase with the increase of reaction time and temperature. Under the same working conditions mentioned above, adding a certain amount of inorganic acid for hydrothermal treatment, the humus content is slightly higher than that of the group without acid. The humification rate reaches 0.45, which is 0.3 times higher than that of the group without acid. It shows that hydrothermal treatment can promote the formation of humus to a certain extent. Compared with prolonging the reaction time, the increase of temperature has a more obvious effect on the increase of humification rate.
Keywords: Aquatic waste; Hydrothermal treatment; Humification; Humus.

1 Introduction

Hydrothermal treatment can save the process of dehydration or drying, which is suitable for the treatment of biomass with high water content [1]. Compared with extracting a single product, the hydrothermal products of aquatic wastes after hydrothermal treatment can be fully used in agriculture and forestry, which has a better prospect. Some studies have shown that after inorganic acid-hydrothermal treatment, due to the promotion of hydrogen ions in the whole process, the humus content and humification rate have been significantly improved . Therefore, this paper focuses on aquatic wastes, and investigates the effects of reaction temperature, reaction time and inorganic weak acid on the formation and transformation of humus in hydrothermal process. Investigate the acid-base behavior in hydrothermal processing of aquatic wastes.FTIR scan to assess humification process of acid-hydrothermal

* Corresponding author: zhanghj@ctbu.edu.cn

(acid-HT) after treatment. Study the relationships between water-soluble carbohydrate and phenol fractions (with more attention to the polyaromatic hydrocarbons concentration) and the humification indices of such aquatic wastes during the treatment process.The aim is to study the humification process and mechanism of aquatic wastes in hydrothermal process, and to provide guidance for subsequent research and application.

2 Materials and methods

The experimental raw materials are aquatic wastes (such as fish head, viscera, skin,). Rinse the raw materials repeatedly with deionized water. After cleaning, put the raw material samples into a cooking machine and smash them, and then put them into sealed bags for later use. The remaining samples are frozen (-4°C) for storage. The principal components of raw material samples are determined in Table 1. Table 1 – Compare these findings (especially humic acids) with the change of structural characterization of the hydrothermal-humification of aquatic wastes.

Table 1. Characteristics of the Dry Aquatic Waste.

Moisture content %	Organic content %	Humification content %	Humification riton	pH
80.5±0.13	85±1.83	8.25±0.27	0.096-0.135	6.75

Note: The content of organic matter is the ratio of the mass of organic matter to the total mass of dry basis; Humus content: the ratio of humus mass to total mass on dry basis; Humification rate is the ratio of humic acid content to fulvic acid content.

The hydrothermal treatment device used in the experiment is a 1L autoclave (FCF-1, Gongyi City Yingyu Boyan Instrument and Equipment Factory), and the centrifugal equipment is (TG165, Shanghai Luxiang Instrument). Freeze-drying (LGJ-18A, Beijing Sihuan Scientific Instrument); Total organic carbon analyzer (TOC-L-CPH, Shimadzu Instrument); Fourier transform infrared spectrometer (IRPrestige-21, Shimadzu Instrument).

The conditions of hydrothermal treatment are reaction temperature of 165°C, 180°C and 195°C respectively. The treatment time are 20, 40, 60 and 80 minutes, and acid-hydrothermal treatment (Acid-HT) with 0.01% sulfuric acid. The heated sample is centrifugally dehydrated, dried in a drying oven at 40°C for 1 hour and dried at -47°C for 24 hours in low temperature vacuum (LGJ-18A type), and then analyzed. The content of organic matter was determined by potassium dichromate method in NY525-2012 standard. Humus is determined by sodium pyrophosphate extraction step by step, and its total carbon content is determined by potassium dichromate volumetric method, and the content of humic acid and fulvic acid is measured. In this paper, the humic rate is defined as the ratio of humic acid content to fulvic acid content.

3 Results and discussion

3.1 Changes of pH by hydrothermal treatment

Figure 1 shows the change of pH of products under different hydrothermal conditions. In the process of hydrothermal reaction, the macromolecular organic matter in the raw materials of aquatic waste samples breaks the bond, and some acidic groups such as carbon group, carboxyl group, aldehyde group and phenolic hydroxyl group are generated. At the same time, during the reaction, a part of the generated organic acids enter the solution, which makes the solution after the reaction acidic. It can be seen from Figure 1 that when the reaction time reaches 20 minutes, the pH of the solution is already acidic. However, when the reaction time of acid-hydrothermal treatment (acid-HT) reached 60 minutes, the pH of the solution slightly increased after the reaction, which indicated that some organic acids in the product were decomposed at this time. This made the pH of the solution increase. On the whole, however, with the increase of temperature and the extension of time, the pH of the solution after the reaction decreased to varying degrees, which indicated that with the intensification of hydrothermal degree, the primary hydrolysis reaction of organic matter became more intense. Easily degradable organic matter was decomposed under the promotion of hydrogen ions of water self-ionization, and organic acids were generated. At the same time, the generated organic acids promoted the hydrolysis reaction, and the pH of the solution after the reaction showed a downward trend. This downward trend is especially obvious when the reaction temperature is 195°C.

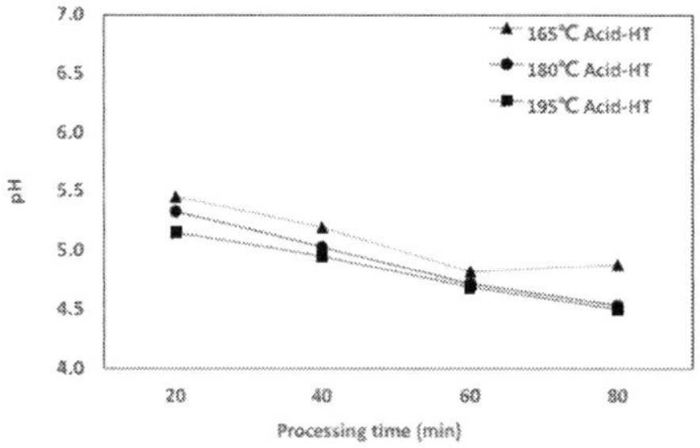

Fig .1. pH of products under different hydrothermal conditions.

3.2 Changes of organic matter by hydrothermal treatment

Figure 2 shows the change of organic matter under different hydrothermal conditions. From the figure, it can be seen that in the initial hydrolysis stage of hydrothermal reaction, there is obvious loss of organic matter. The reason is that in this stage, acromolecules are decomposed into small molecules and dissolved in the liquid phase. With the extension of treatment time, the hydrothermal reaction went to the second stage, and some small molecules condensed with aromatic compounds to generate various new organic matters[2]. Therefore, the content of organic matter began to show an upward trend again. It can also be seen from Figure 2 that in acid-hydrothermal treatment (acid-HT) at 165°C, when the hydrothermal treatment time is 40-60 minutes, the organic matter content always shows a slow downward trend, while at 180°C and 195°C, the organic matter content begins to show an upward trend after 40 minutes of hydrothermal reaction. This shows that the increase of temperature can accelerate the decomposition reaction at the initial stage of hydrothermal treatment. It can also be found from the figure that there is a certain loss of organic matter after hydrothermal treatment, but the loss is relatively small, and the content of organic matter reaches 79.1% at 165°C and 20min, which is 45% higher than the national standard of organic fertilizer agriculture.

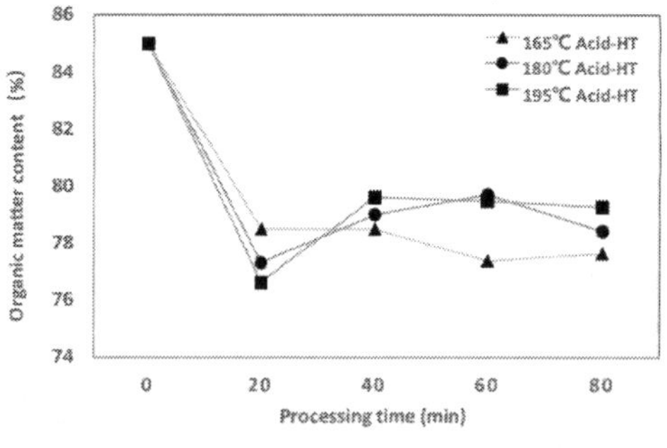

Fig. 2. Organic matter content of products under different hydrothermal conditions.

3.3 Effect of hydrothermal treatment on humification process

Figure 3 shows the influence of hydrothermal treatment on humification process. In this paper, the humification rate is defined as the ratio of humic acid content to fulvic acid content. Although there is no corresponding evaluation standard for humification rate in hydrothermal treatment, hydrothermal products are similar to compost products, this paper evaluates the maturity of hydrothermal process with reference to the evaluation standard of compost. It is generally believed that when the humification rate is greater than > 0.8, the compost can be considered as mature [3]. The humic rate of raw material samples is 0.096, and the degree of maturity is extremely low. Acid-hydrothermal treatment (acid-HT) has a humification rate of 0.35 at 180°C for 60min. With the increase of temperature to 195°C, the humification rate of hydrothermal products increases more obviously. When the reaction time reaches 60min and 80min, the humification rate is 0.44 and 0.45 respectively. Figure 3 shows that under six hydrothermal conditions, the humification rate does not meet the requirement of the evaluation standard that the humification degree is greater than 0.8, but it is increased by about 2-4 times compared with the humification rate of raw materials, which shows that hydrothermal treatment can promote the formation of humus to some extent.

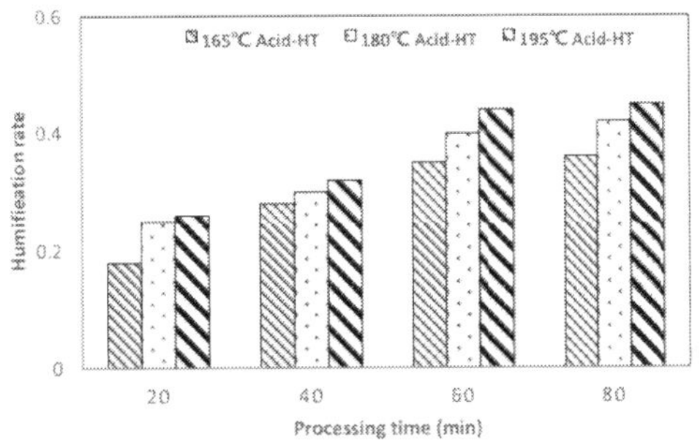

Fig. 3. Humifieation rate of products under different hydrothermal conditions.

3.4 Infrared spectrum analysis

By searching the spectrogram [4], it is recognized that the infrared spectrogram drawn by the acid-hydrothermal (acid-HT) at 165°C for 80 minutes belongs to the following absorption peaks: one absorption peak of aromatic substances (benzene skeleton stretching vibration at 1516.40cm-1) and two absorption peaks of sugar substances (OH stretching vibration of sugar C-OH at 448.79 cm^{-1}; stretching vibration of sugar C-OH at 1016.56cm^{-1}).

Through the comparison of spectrogram retrieval, it is recognized that the absorption peaks of each region in the infrared spectrogram drawn by the acid-hydrothermal (acid-HT) at 195°C for 80 minutes belong to the following: four absorption peaks of aromatic substances (Aromatic aldehyde stretching vibration at 1,726.65 cm-1; aromatic C=C telescopic vibration at 1,616.49 cm-1; benzene-like skeleton stretching vibration at 1,508.41 cm-1; symmetric stretching vibration of aromatic acid ester C-O-C at 1036.63cm-1),1 carbohydrate absorption peak (OH stretching vibration of sugar C-OH at 3446.79cm-1),1 absorption peak of phenolic alcohols (-OH Telescopic Vibration of Phenol Hydroxyl at 3417.79cm-1).

From the above analysis, it can be seen that the products under hydrothermal conditions at 195°C for 80 minutes contain more aromatic functional groups than those under hydrothermal conditions at 165°C for 80 minutes. This shows that with the deepening of the reaction degree, the formation of macromolecular humic acid becomes more intense, so the aromatic compounds in the product increase. This also shows that hydrothermal reaction can

 https://doi.org/10.1051/e3sconf/202234101012

promote the formation of humus to a certain extent, and aquatic wastes can gradually mature with the increase of temperature in hydrothermal reaction. which is consistent with the above inference results (3.3).

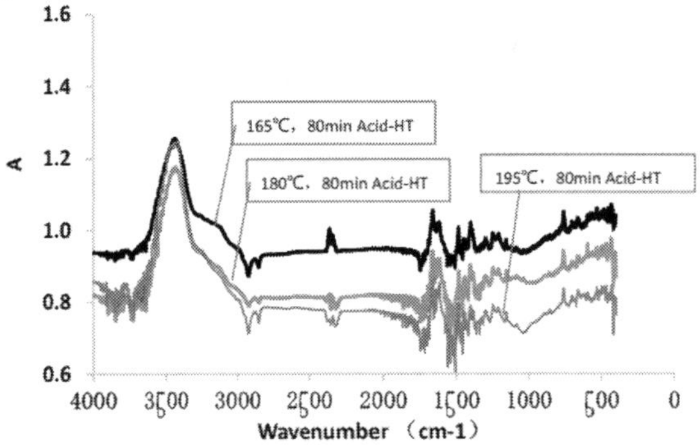

Fig. 4. The infrared spectra different hydrothermal conditions.

4 Conclusion

In the actual working condition, the humification rate of acid-water heat treated samples at 195°C for 80min is 0.45, which only reaches 50% of the requirement of the evaluation standard that the degree of humification is > 0.8, but compared with the humification rate of raw materials, it is increased by about 2-4 times, indicating that hydrothermal treatment can promote the formation of humus to a certain extent. The experimental results show that, compared with prolonging the reaction time, the increase of temperature has a more obvious effect on the increase of humification rate.

Funding: National Natural Science Foundation of China [Project No.51808079], the natural science foundation of Chongqing [Project No. cstc2017jcyjAX0470 and cstc2020jscx-lyggX0001] and Chongqing Technology and Business University Scientific Research Foundation [Project No. 670101379 and CQCM-2016-07].

References

1. Dewatering, Dehydration and drying of sludge. Anaerobic Sewage Treatment: Optimization of Process and Physical Design of Anaerobic and Complementary Processes, 2019, pp.257-291.
2. Evaluation of physicochemical characteristics and health risk of polycyclic aromatic hydrocarbons in borehole waters around automobile workshops in Southeastern Nigeria. Groundwater for Sustainable Development 14, 100615. 2021.
3. Mahmoud I B, Medhioub M, Rigane H, et al. Evolution of humicfractions in calcimagnesic soils and brown isohumic soils amended with compost and manure[J]. Compost Science&Utilization, 2007,15(4):253
4. S Wenger Shi-Pu, Fourier transform infrared spectrum analysis, Chemical Industry Press (CIP) 2004.

Study on the antioxidant ability of procyanidins and their complexes

Di Lu, *Lijuan* Wang, *Wu* Zhang, *Baohui* Guo, and *Yuguang* Lv*
College of Pharmacy, Jiamusi University, Jiamusi, China

Abstract. Due to the excellent antioxidant capacity of proanthocyanidins, combined with the antioxidant mechanism of anthocyanins, it inspired us to study the antioxidant capacity of proanthocyanidins and their complexes. In this article, we conducted antioxidant tests on proanthocyanidin-bovine serum albumin and procyanidin-metal complexes, using DPPH and ABTS methods to determine the antioxidant capacity of procyanidins and their complexes. The results show that compared with proanthocyanidins, the proanthocyanidin complex has a certain degree of improvement in the ability to scavenge free radicals. After being combined with metal ions, the ability to scavenge free radicals is significantly improved. When the ratio of proanthocyanidin B2 to Fe^{3+} is 2:1, the ability to scavenge free radicals is significantly enhanced.
Keyword: Proantho Cyanidins; Anti-oxidation; Free radicals.

Preface

Proanthocyanidins are a general term for a large class of polyphenolic compounds that are widely present in plants. They have strong antioxidant and free radical scavenging effects, and can effectively eliminate hydroxyl radicals and superoxide anion radicals. They also promote the synthesis of arachidonic acid and phosphoric acid, which protect lipids from peroxidative damage in the process. They are a powerful metal chelator that chelates metal ions to form inert compounds in the body; protects and stabilizes vitamin C and aids in vitamin C absorption and utilization.

Oxidative stress is related to the pathogenesis of human degenerative diseases such as diabetes, cancer and cardiovascular disease [1-3]. The antioxidant mechanism of anthocyanins is mainly manifested in reducing the accumulation of reactive oxygen species (ROS), scavenging free radicals, and reducing DNA damage.

ROS is an oxygen-containing intermediate metabolite, including oxygen free radicals and non-free radicals. Oxygen free radicals include superoxide anion radicals (O_2^-), three-electron reduction products of oxygen, hydroxyl radicals (OH), peroxy radicals (ROO) and nitric oxide (NO), etc., which are not free radicals. Including the two-electron reduction products of oxygen, hydrogen peroxide (H_2O_2), peroxynitrite ($ONOO^-$), and singlet oxygen ($^1O_2$), etc., which are converted into new free radicals through the oxidation of other components[4-7].

Naturally occurring anthocyanin aglycones and their glycosides have an effect on

*Corresponding author: yuguanglv@163.com

1,1-diphenyl-2-picrylhydrazyl (DPPH), OH, 2-diazo-bis(3-ethyl) -Benzothiazole-6-sulfonic acid) diammonium salt free radical [2,2'- azinobis (3-ethylbenzothiazoline-6-sulfonic acid ammonium salt), $ABTS^+$], etc., have scavenging ability, and have a slight scavenging ability for DPPH and OH .Compared with the same concentration of vitamin C (vitamin C, VC), its ability to clear $ABTS^+$ is slightly stronger.[8].

1 Experimental method

In this study, two antioxidant detection methods were selected: DPPH method and ABTS method.

2 Experimental result

2.1 Results of the ability of proanthocyanidins to clear DPPH under different pH values

2.1.1 The clearance rate of proanthocyanidin samples in absolute ethanol

According to the sample amount of proanthocyanidin standard solution , the test solutions with mass concentrations of 0.1 mg/mL, 0.2 mg/mL, 0.3 mg/mL, 0.4 mg/mL and 0.5 mg/mL can be obtained. Add sequence according to different samples, the experimental results obtained in parallel 2 groups are: Under the corresponding concentration, the clearance rate of proanthocyanidin in absolute ethanol solution is 91.95%, 91.35%, 92.80%, 92.65% and 93.30%.

2.1.2 Results of the ability of proanthocyanidin samples to remove DPPH under different pH values

According to the sample volume of the proanthocyanidin standard solution , the mass concentrations of 0.1 mg/mL, 0.2 mg/mL, 0.3 mg/mL, 0.4 mg/mL and 0.5 mg/mL can be obtained at different pH values. Test the solution, add sequence according to different samples, and the experimental results obtained in parallel for 2 groups are: at the corresponding concentration, the clearance rate of proanthocyanidin absolute ethanol solution after adding different pH buffer solutions is changed, as shown in Table 1.

Table 1. The clearance rate of proanthocyanidins of various mass concentrations at different p H.

Concentration	0.1mg/mL	0.2mg/mL	0.3mg/mL	0.4mg/mL	0.5mg/mL	pH
	95.55	95.80	95.75	95.55	95.60	3
	92.90	92.45	92.85	92.55	92.95	5
Clearance rate	48.50	50.60	56.10	57.45	56.00	7
(%)	37.45	39.50	44.25	45.05	46.80	9
	36.60	41.40	42.70	43.45	43.70	11
	-16.80	-14.45	-14.75	-14.95	-13.50	13

According to the experimental results, the ability of proanthocyanidins to clear DPPH under different pH values was drawn as shown in Figure 1.

In the DPPH removal experiment, it can be seen that when proanthocyanidin is under acidic conditions, its clearance rate is slightly higher than that of proanthocyanidin ethanol solution. When proanthocyanidin is under neutral and alkaline conditions, its clearance rate is lower than that of proanthocyanidin-ethanol solution. And as the alkalinity increases, the clearance rate decreases.

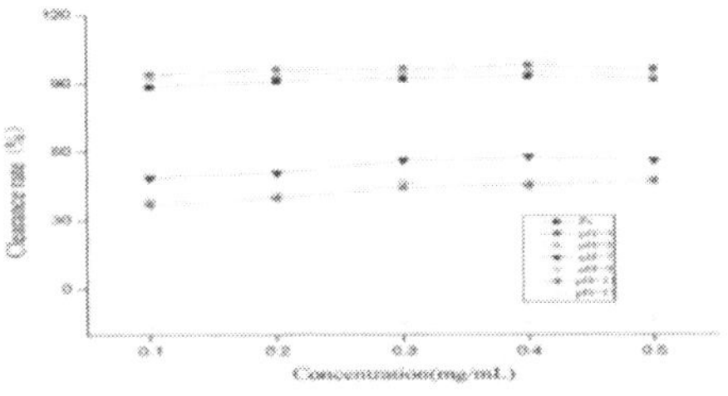

Fig. 1. The ability of proanthocyanidins to clear DPPH under different pH value.

2.3 Study on the Antioxidant Activity of Procyanidins in vitro

2.3.1 Determination of the ability of proanthocyanidin complexes to scavenge $ABTS^+$

In the experiment, in order to evaluate the anti-oxidation ability of proanthocyanidin complexes, a scavenging experiment was carried out, and the anti-oxidation comparison analysis with Vc was carried out. The experimental results are shown in Figure 2.

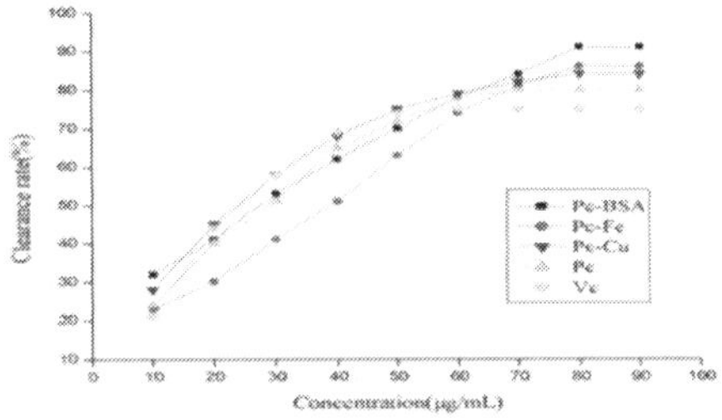

Fig. 2. Proanthocyanidins and its complexes, Vc on the removal of ABTS+.

Figures 2 show that proanthocyanidins, their complexes, and Vc have a scavenging effect on $ABTS^+$, and the ability to scavenging DPPH gradually becomes stronger as the concentration increases.

Based on the linear regression simulation of procyanidin-iron, proanthocyanidin-copper, proanthocyanidin-bovine serum albumin, proanthocyanidin and Vc clearance $ABTS^+$antioxidant curve in Figure 2, the correlation coefficient R^2 and the linear simulation equation are obtained, and their antioxidant capacity is compared and analyzed.

Table 2. Analysis of anti-oxidation results of proanthocyanidins and their complexes and Vc.

sample	Linear regression equation	R^2
A: Proantho Cyanidins	$Y=-0.00126X^2+1.9547X+5.5952$	0.9975
B: Procyanidins-Bovine Serum Protein	$Y=-0.0055X^2+1.3231X+18.143$	0.9969
C: Proanthocyanidin-Iron	$Y=-0.0054X^2+1.4095X+6$	0.9876
D: Procyanidin-copper	$Y=-0.0117X^2+1.8355X+12.238$	0.9966
Vc	$Y=-0.0159X^2+2.1737X+4.619$	0.9749

It can be seen from the analysis of Table 1-2 that the order of $ABTS^+$ scavenging ability of the samples is: procyanidin-bovine serum albumin>procyanidin-iron>procyanidin-copper>procyanidin>Vc, and the concentration has a good linear relationship with the clearance rate.

2.3.2 Determination of the ability of proanthocyanidin complexes to scavenge DPPH

The scavenging ability of primary free radicals was determined by the DPPH antioxidant activity method. The scavenging ability of proanthocyanidin complex to DPPH was determined, and the results are shown in Figure 3.

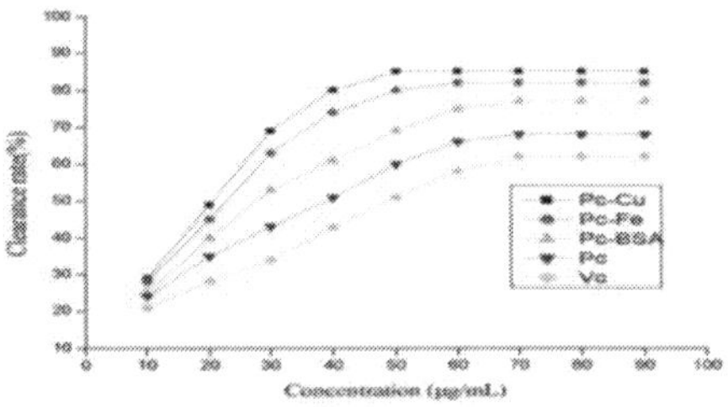

Fig. 3. Proanthocyanidins and its complexes, Vc on the removal of DPPH.

Figure 3 shows that proanthocyanidin and its complexes and Vc have a scavenging effect on DPPH, and the ability to scavenging DPPH gradually becomes stronger as the concentration increases.

Table 3. Analysis of anti-oxidation results of proanthocyanidins and their complexes and Vc.

sample	Linear regression equation	R^2
A: Proantho Cyanidins	$Y=-0.0084X^2+1.4108X+908571$	0.9945
B: Procyanidins-Bovine Serum Protein	$Y=-0.0114X^2+1.7789X+808333$	0.9986
C: Proanthocyanidin-Iron	$Y=-0.0156X^2+2.1834X+8.9524$	0.9895
D: Procyanidin-copper	$Y=-0.0174X^2+2.3542X+9.8095$	0.9717
Vc	$Y=-0.006X^2+1.158X+7.7619$	0.9847

It can be seen from the analysis of Tables 3 that the order of the clearance capacity of the samples is: procyanidin-copper>procyanidin-iron>procyanidin-bovine serum albumin>procyanidin>Vc, and the concentration has a good linear relationship with the clearance rate.

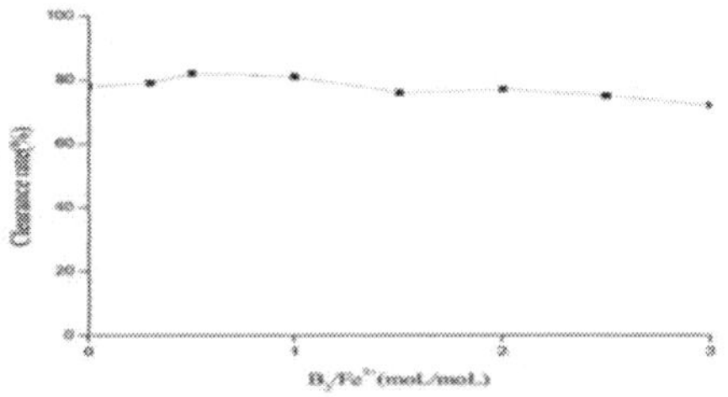

Fig. 4. The scavenging effect of different ratios of proanthocyanidins B2 and Fe^{3+} on DPPH free radicals.

It can be seen from Figures 4 that adding a small amount of metal ions to proanthocyanidin B2 has a stronger ability to scavenge free radicals. When the molar ratio of proanthocyanidin B2 to Fe^{3+} is 2:1, the ability to scavenge free radicals is the highest.

4 Conclusion

In this subject, the antioxidant capacity of procyanidins and their complexes was determined. The results showed that the antioxidant capacity of procyanidin complexes was improved compared with that of procyanidins. After combining with metal ions, the antioxidant capacity was significantly improved.

References

1. AZUMA A, YAHUSHIJI H, SATO A. Postharvest light irradiation and appropriate temperature treatment increase anthocyanin accumulation in grape berry skin [J].Postharvest Biol Technol,2019,147(1): 89-99.
2. YOU L, SUI QQ, ZHAO YX, et al. Recent progress in structural modification and physiological activity of anthocyanins [J]. Food Sci.2019,40(11):351-359.
3. ANDRADE TA, HAMERSKI F. LOPEZ FDE, et al. Ultrasound-assisted pressurized liquid extraction of anthocyanins from Aronia melanocarpa pomace [J]. Sep Purif Technol,2021,267(12):119290.
4. KUMAR M, DAHUJA A, SACHDEV A, et al. Evaluation of enzyme and microwave-assisted conditions on extraction of anthocyanins and total phenolics from black soybean (Glycine max L.) seed coat [J]. Int J Biol Macromol,2019,135(8): 1070-1081.
5. FERREIRA FL, MINUZZI NM, RODRIGUES RF, et al. Citric acid water-based solution for blueberry bagasse anthocyanins recovery Optimization and comparisons with microwave-assisted extraction (M AE)[J].LWT,2020,133(11): 110064.
6. ODABAS HI, KOCA I. Simultaneous separation and preliminary purification of anthocyanins from Rosa pimpinellifolia L. fruits by microwave assisted aqueous two-phase extraction [J]. Food Bioprod Process,2021,125(1):170-180.
7. WANG YW, LUAN GX, ZHOU W, et al. Subcritical water extraction, UPLC-Triple-TOF/MS analysis and antioxidant activity of anthocyanins from Lycium ruthenicum Murr [J]. Food Chem,2018,249:119-126.
8. LIMA AS, OLIVEIRA BS, SHABUDIN SV. et al. Purification of anthocyanins from grape pomace by centrifugal partition chromatography[J].J Mol Liq.2021.326(3):115324.

Remote sensing comprehensive evaluation of land degradation based on relative rain use efficiency of vegetation

Yu Xing*, *Jinzhong* Yang, *Yaqing* Sun, *Cunhao* Jiang, *and Dong* Chen

China Aero Geophysical Survey and Remote Sensing Center for Natural Resources, 100083 Beijing, China

Abstract. China is one of the countries with serious land degradation problems in the world. Land degradation directly threatens the food security of our country and determines the sustainable development of social economy. In this paper, the potential evapotranspiration and moisture index are calculated to divide the climatic zone of China. Based on the NPP and precipitation data, the national rRUE is obtained by the edge function. With the climate "relative benchmark", the rules of threshold analysis is used to comprehensively evaluate the land degradation in China. The results show that: excellent land is concentrated in northeast China, western Xinjiang and eastern Qinghai-Tibet Plateau; mildly degraded land is widely and sporadically distributed in the eastern plain of China; moderately degraded land and severely degraded land are concentrated in western Inner Mongolia, southern Xinjiang and northern Qinghai-Xizang Plateau.

1 Forward

Under the strong interference of human activities, land resources are faced with land degradation problems such as desertification, soil erosion, soil pollution, grassland degradation and sharp reduction of wetlands. The qualitative change process of land degradation is the transformation process from land resource types with high productivity or ecological service function to sandy land, bare land, bare rock gravel land, saline land and non-agricultural construction land which are difficult to use or have very low productivity[1]. From an ecological point of view, land degradation is the deterioration of plant growth conditions and the decline of land productivity. There is a very close relationship between vegetation cover and land degradation. Deforestation, overgrazing and vegetation destruction are the direct causes of land degradation. The study shows that there is a negative exponential relationship between vegetation cover and land degradation, and the correlation coefficient is 0.8[2]. Vegetation degradation is not only the cause of land degradation, but also the result of land degradation. Rain use efficiency of vegetation (RUE) is a comprehensive index to describe the productivity and health status of vegetation in ecosystem. Yin selected the RUE with high time resolution from remote sensing inversion

*Corresponding author: 63216429@qq.com

as the evaluation index, and used the time series analysis method to evaluate the development trend of desertification in Inner Mongolia Autonomous region[3]. Wang monitored and evaluated the land degradation in Beijing-Tianjin wind-blown sand source area based on net primary productivity(NPP) and RUE[4]. Gao also discussed the remote sensing evaluation method of desertification based on RUE[5]. However, there are few studies on large-scale evaluation of land degradation. Based on NPP and climatic data, this paper selects the index of relative rain use efficiency of vegetation (rRUE), combined with the significance test of the correlation coefficient of NPP with time, quantitatively evaluates and rapidly monitors the situation of land degradation in China.

2 Methods

2.1 Technical route

First of all, the RUE is calculated from the NPP and precipitation data. Secondly, the potential evapotranspiration and moisture index are calculated by using meteorological data[6-7]. And then, with reference to relevant standards, the land of China is divided into 18 climate gradient zones. Next, based on the climate gradient zone, the edge function model is established, from which the relative rain use efficiency of vegetation is obtained, and the threshold analysis method is used to classify the land production capacity of rRUE. Finally, on the basis of considering the "relative benchmark" of climate, the significance of the correlation coefficient of NPP with time change is used. The annual average and annual maximum of rRUE are used to comprehensively evaluate the land degradation situation in China in the past 10 years. The research route is shown in Fig. 1.

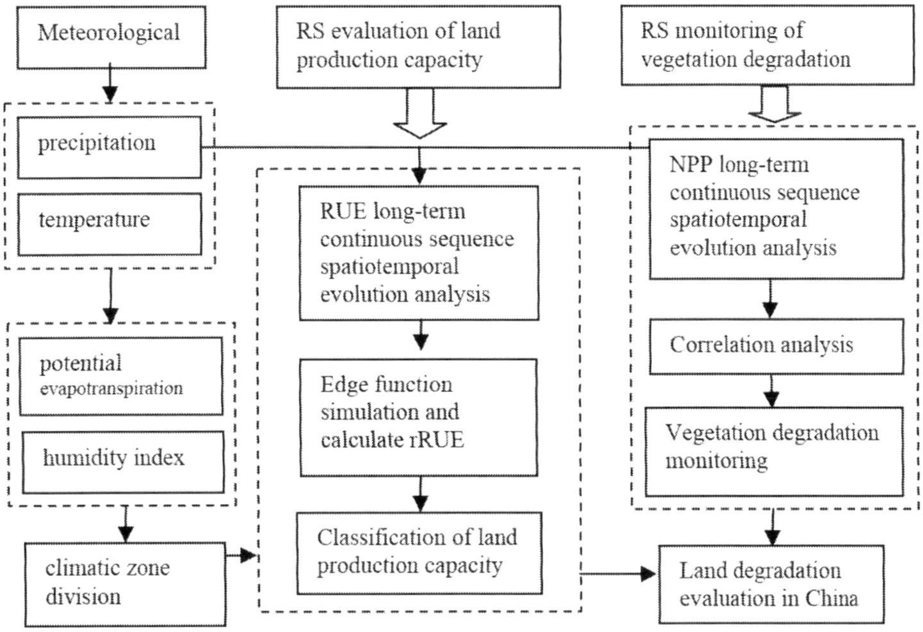

Fig.1. Technical route.

2.2 Comprehensive evaluation

The mean value of RUE (RUEmean) represents the long-term vegetation productivity and production capacity of the ecosystem, while maximum value of RUE (RUEmax) represents the resilience of the ecosystem after disturbance, especially the interference of the interannual variation of annual precipitation [4]. The threshold analysis is used to classify the land production capacity based on the mean value of rRUE (rRUEmean) and the maximum value of rRUE (rRUEmax). The evaluation method is as follows:

• Severely degraded land: when *rRUE*mean < 0 and *rRUE*max < 0 in the study area, it means that the ecosystem production capacity in this area is lower than the minimum of its potential production capacity, that is, the land in this area has almost no production capacity, and the resilience of ecosystem anti-interference is also lower than the minimum of its potential elasticity, that is, the land in this area is almost unable to restore its productivity and does not have ecological recovery elasticity. These areas are defined as extremely poor production capacity in the study.

• Moderately degraded land: when *rRUE*mean < 0 and 0 ≤ *rRUE*max ≤ 1 in the study area, it means that the ecosystem production capacity in this area is lower than the minimum of its potential production capacity, that is, the land production capacity in this area is very low, but the regional ecosystem has normal anti-interference resilience, and the land in this area can also restore its productivity. When 0 ≤ *rRUE*mean ≤ 1 and *rRUE*max < 0 in the study area, it means that the ecosystem production capacity in this area is within the normal potential range, and the land production capacity in this area is at the annual average level. But the resilience of ecosystem anti-interference in this area is lower than its minimum potential elasticity, that is, the anti-interference ability of land in this area is low and does not have ecological restoration elasticity.

• Mildly degraded land: when 0 ≤ *rRUE*mean ≤ 1 and 0 ≤ *rRUE*max ≤ 1 and *R*npp< -0.5214 in the study area, it means that the ecosystem production capacity in these areas is within the normal potential range, the land production capacity in this area is at the annual average level, and the ecosystem has normal anti-interference recovery elasticity, the land in this area has the potential to restore productivity. However, *R*npp<-0.5214 indicates that the vegetation is degraded. So, these areas are defined as land with mildly degraded land. (*R*npp represents the correlation coefficient of NPP index and wetness index with time, and 0.5214 is the critical value of significance test of correlation coefficient at 0.1confidence level.)

• Good land: when 0 ≤ *rRUE*mean ≤ 1 and 0 ≤ *rRUE*max ≤ 1 and *R*npp≥ -0.5214 in the study area, it means that the ecosystem production capacity in these areas is within the normal potential range, the land production capacity in this area is at the annual average level, and the ecosystem has normal anti-interference recovery elasticity, the land in this area has the potential to restore productivity. And *R*npp ≥-0.5214 indicates that the vegetation is restored. So, these areas are defined as land with Good land.

• Excellent land: when *rRUE*mean > 1 and 0 ≤ *rRUE*max ≤ 1 in the study area, it means that the ecosystem production capacity of these areas is very strong, and the ecosystem has normal anti-interference resilience, and the land in this area has the potential to restore productivity. When 0 ≤ *rRUE*mean ≤ 1 and *rRUE*max > 1 in the study area, it means that the ecosystem production capacity in these areas is within the normal potential range, the land production capacity in this area is at the annual average level, and the ecosystem has a strong anti-interference resilience. When *rRUE*mean > 1 and *rRUE*max > 1 in the study area, it means that the ecosystem production capacity of these areas is very strong, and the ecosystem has a strong anti-interference recovery elasticity. This area is defined as land with excellent production capacity.

• Abnormal: when the study area (*rRUE*mean < 0 and *rRUE*max > 1) or (*rRUE*mean > 1 and *rRUE*max < 0), it means that one index of rRUEmean and rRUEmax is less than 0, while the other index is greater than 1. Thus, it is defined as an anomaly area.

According to the above threshold analysis, combined with the correlation between NPP and MI and time, the rules of comprehensive evaluation of land degradation are summarized, as shown in Table 1. *R*npp represents the correlation coefficient of NPP index and wetness index with time, and 0.5214 is the critical value of significance test of correlation coefficient at 0.1confidence level.

3 Results

The annual average temperature, precipitation, potential evapotranspiration, NPP and RUE data are obtained or calculated, and the edge function fitting equations of RUEmean and RUEmax are established. The results are shown in Table 1.

Table 1. RUE upper and RUE lower bound edge function models based on MI data.

y	x	Edge function model
$RUEmean_{0.05}$	$MImean$	$y = 0.0582x^4 - 0.3505x^3 + 0.9242x^2 - 1.1621x + 0.5539$
$RUEmean_{0.95}$	$MImean$	$y = -1.3533x^6 + 12.621x^5 - 46.595x^4 + 86.198x^3 - 81.714x^2 + 32.773x + 2.3314$
$RUEmax_{0.05}$	$MImax$	$y = -0.0313x^6 + 0.1013x^5 + 0.304x^4 - 1.5361x^3 + 2.0817x^2 - 1.2363x + 0.3768$
$RUEmax_{0.95}$	$MImax$	$y = -0.0343x^6 + 1.3229x^5 - 10.011x^4 + 31.39x^3 - 45.935x^2 + 27.314x + 1.1375$

The annual rRUEmean and annual rRUEmax are calculated according to the formula:

$$rRUE=(RUE-RUE_{0.05})/(RUE_{0.95}-RUE_{0.05}). \quad (1)$$

According to the rules of comprehensive evaluation of land degradation, the evaluation results of land degradation in China are obtained (Fig. 2).

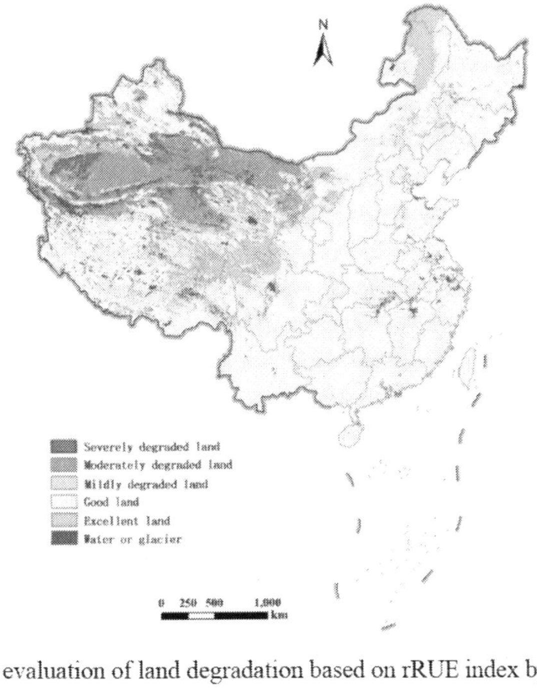

Fig. 2. Comprehensive evaluation of land degradation based on rRUE index by remote sensing.

The total area of degraded land in China is about 800000 km^2, of which the areas of mild degradation, moderate degradation and severe degradation are about 460000 km^2,

140000 km^2 and 200000 km^2 respectively, excellent land about 530000 km^2 and good land about 8.17 million km^2.

4 Conclusion

Based on the analysis of national regional climate and NPP data in recent 10 years, a remote sensing comprehensive evaluation map of land degradation in China is obtained based on NPP change trend and rRUE index. The study shows that excellent land is concentrated in northeast China, western Xinjiang and eastern Qinghai-Tibet Plateau; mildly degraded land is widely and sporadically distributed in the eastern plain of China; moderately degraded land and severely degraded land are concentrated in western Inner Mongolia, southern Xinjiang and northern Qinghai-Xizang Plateau.

Acknowledgements

This paper is supported by Programs of China Geological Survey (Remote sensing monitoring of national mining exploration and ecological spatial in key areas).

References

1. K.F. Zhang, X.W. Li, D.X. Zhang, J. F. Peng, J. Chen, Z. R. Yu, Environmental Science, **27**,1244 (2006)
2. G.N. Sun, M.H. Wang, Journal of Arid Land Resources and Environment, **22**,140 (2008)
3. H. Yin, Z.G. Li, Y.L. Wang, F. Cai, Acta Geographica Sinica,**66**,653 (2011)
4. H.Y. Wang. Assessment and monitoring of land degradation using NPP and vegetation rain use efficiency,*Chinese Academy of Forestry* (2013)
5. Z.H. Gao, Z.Y. Li, G.D. Ding, L. Y. Li, Science of Soil and Water Conservation, **3**, 37 (2005)
6. Thornthwaite, Geographical Review, **38**,55(1948)
7. S.W. Zhuang, H.C. Zuo, P.C. Ren, G. J. Xiong, B. D. Li, W. C. Dong, L. Y. Wang, Climatic and Environmental Research, **18**,617(2013)

Construction of evaluation system and evaluation model of urban environmental engineering geological problems —— a case study of Jinan City

Yunfeng Zhang [1,2], *Ana* Wu [3], *Jianjun* Shen [4], and *Qing* Cao[4*]

[1] Shandong Provincial Geo-mineral Engineering Exploration Instiute, Jinan ,250014, China
[2] Shandong Fifth Institute of Geology and Mineral Exploration, Taian, 250013, China
[3] Shandong GEO-Surveying & Mapping Institute, Jinan, 250002, China
[4] College of Chemical Engineering and Safety, Binzhou University, Binzhou, 256600, China.

Abstract. In order to further study the problems of urban engineering geological environment, taking Jinan as an example, the environmental engineering geological problems are divided into four aspects: primary environmental engineering geological problems, secondary environmental engineering geological problems, the influence of human activities and the restriction of spring water on engineering construction. A multi-level, multi-objective and multi factor evaluation index system suitable for environmental engineering geology is established. The mathematical model is established by using analytic hierarchy process and fuzzy comprehensive evaluation method to evaluate the primary environmental engineering geological problems, secondary environmental engineering geological problems, the impact of human activities and the restrictive effect of spring water on engineering construction. The mathematical model is established by using Nemero index method to comprehensively evaluate the environmental engineering geological problems in Jinan, and the suitability zoning of engineering construction is carried out.

1 Introduction

Environmental engineering geology is a new discipline. Artificial and natural environments jointly shape cities, and urban construction and geological environment promote and restrict each other [1]. Therefore, in urban engineering geology, it is necessary to predict all types of geological engineering problems in cities and suggest corresponding prevention and control strategies [2].

The study of urban engineering geological environment in the modern sense began in the late 1960s when engineering geology in urban planning was gradually realised in practice. With the acceleration of urbanisation and rapid development of economic construction in China, in mining engineering, water conservancy and hydropower, urban construction, geological disasters, water and soil pollution, and other aspects of

*Correspongding author: caoqing526@163.com

environmental engineering geology have carried out a series of research, and achieved a lot of research results. Human engineering activities have accelerated the transformation of the natural geological environment, which has become a driving force keeping pace with natural action. In addition, in big cities with dense populations and frequent economic activities, the consequences are often quite serious once geological disasters or environmental accidents occur. They cause great economic losses and social impact, even causing local social instability [3-6]. Therefore, the study of modern urban environmental engineering geology has essential theoretical and practical significance for correctly understanding the law of environmental engineering geology, revealing, reducing, and avoiding the environmental engineering geology problems, and improving the environmental quality of urban engineering geology.

Taking Jinan city as an example, this paper summarizes the environmental engineering geology problems into four aspects: primary environmental engineering geology problems, secondary environmental engineering geology problems, the effect of human activities and the restriction of spring water on engineering construction. A multi - level, multi - objective and multi - factor evaluation index system suitable for environmental engineering geology is established. The factors affecting environmental engineering geology are quantitatively analyzed and evaluated by analytic hierarchy process(AHP) and fuzzy comprehensive evaluation method. The safety of environmental engineering geological problems in Jinan city is divided into districts.

2 Main contents of urban environmental engineering geology research

With the development of society, the impact of human activities on the engineering geological environment has become increasingly evident, and the environmental engineering geological problems caused by this have become increasingly severe and complex [7-18].

(1) Traditional regional engineering geology problems: It includes soft soil, expansive soil, collapsible loess, liquefied sand, and other special soils, as well as deformation failure of deep and large foundation pits.

(2) Geological disasters: It includes collapse, landslide, debris flow, ground subsidence, ground collapse, ground cracks, and geological disasters caused by mine development.

(3) Seismic geology includes crustal stability and tectonic earthquakes formed by active faults.

(4) Geological problems formed by human engineering economic activities: It includes mine tailings pond, garbage storage yard, artificial water storage reservoir, underground space development, and groundwater overexploitation.

(5) Geochemical and environmental problems include water and soil pollution and heavy metal pollution.

(6) Issues related to land governance: It includes soil erosion and land desertification.

3 Evaluation Methods of urban environmental engineering geology

At present, foreign countries have combined sustainable urban development with engineering geology, and the study of engineering geology has leapt from quantitative change to qualitative change [19-25].

With the development of the economy, standardisation of the construction industry, and gradual attention to engineering geology, urban engineering geology developed into a specialised discipline [26-36]. Miao selected the grey clustering evaluation model and fuzzy comprehensive evaluation model, adopted the priority relation ranking method and principal component analysis method to screen and optimise each evaluation index, and conducted comprehensive analysis and evaluation on the geological environment quality [37-39].

This paper used analytic hierarchy process (AHP) and fuzzy combined evaluation method to establish a mathematical model to evaluate the primary environmental engineering geology problems, secondary environmental engineering geology problems, the impact of human activities, and the restricting effect of spring water on engineering construction. Nemerow index method was used to establish a mathematical model to comprehensively evaluate the environmental engineering geological problems in Jinan city, and to partition the engineering construction suitability. The evaluation model is shown in Figure 1. The index classification of karst collapse evaluation is shown in Table 1.

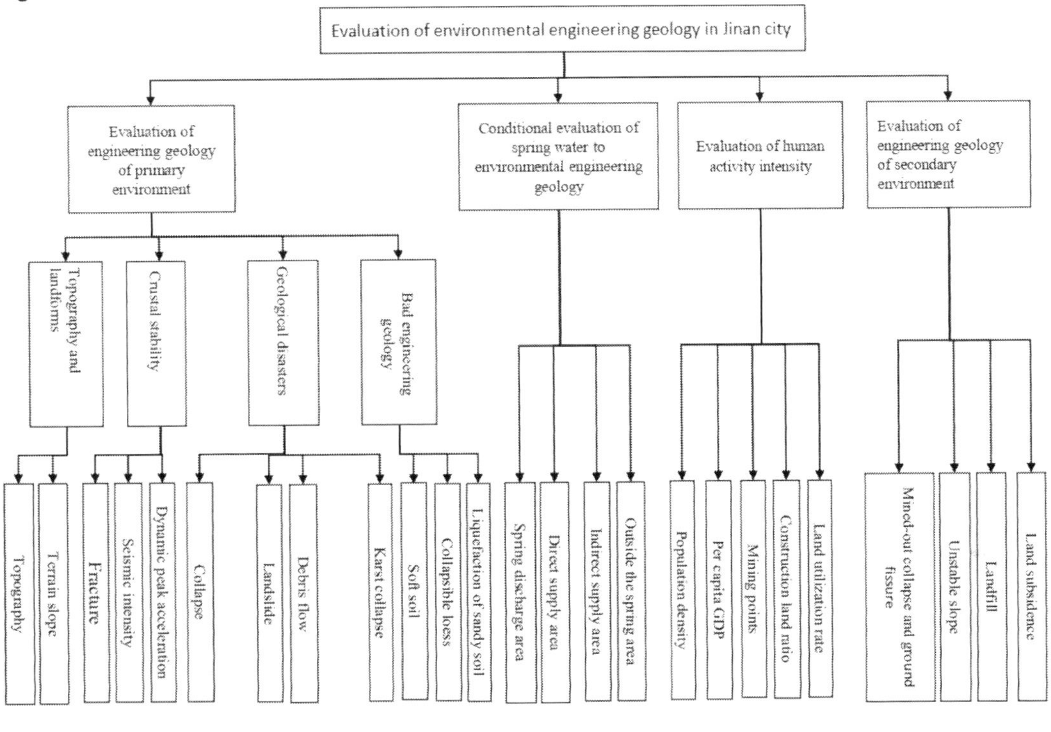

Fig. 1. Evaluation system.

Using the grid division function of the MAPGIS software, the city's grids were named from 1 to 2029 according to the User ID, and a total of 2029 grids are divided. The factors affecting major environmental engineering geological issues are primary environmental engineering geological factors and secondary environmental engineering geological factors. Because the primary environmental engineering geological problems and the secondary environmental engineering geological problems are complementary in affecting the major environmental engineering geological problems of Jinan, the Nemerow index method and the analytic hierarchy process are combined to comprehensively evaluate the two secondary indicators. The results are shown in Figure 2 and Figure 3.

Table 1. Hierarchical classification table of karst collapse evaluation indexes.

Factors of grade four	Factors of grade five	Factors of grade six	Risk level
Karst collapse D_4	Bedrock lithology E13	Dominated by limestone	Ⅰ High risk
		Mixture of limestone and sandstone	Ⅱ Risk

 https://doi.org/10.1051/e3sconf/202234101015

	Dominated by sandstone	Ⅲ Less risk
	Dominated by siltstone	Ⅳ Small risk
Karst development degree E14	Stronger	Ⅰ High risk
	Strong	Ⅱ Risk
	Ordinary	Ⅲ Less risk
	Weak	Ⅳ Small risk
Overburden thickness(m)E15	0-30	Ⅰ High risk
	30-50	Ⅱ Risk
	50-70	Ⅲ Less risk
	>70	Ⅳ Small risk
Groundwater depth(m)E16	0-15	Ⅰ High risk
	15-30	Ⅱ Risk
	30-45	Ⅲ Less risk
	>45	Ⅳ Small risk

Figure 2 shows the risk assessment results of engineering geological problems in the primary environment. Figure 3 shows the risk assessment results of engineering geological problems in the primary environment.The factors affecting major environmental engineering geological issues are primary environmental engineering geological factors and secondary environmental engineering geological factors. Because the primary environmental engineering geological problems and the secondary environmental engineering geological problems are complementary in affecting the major environmental engineering geological problems of Jinan, the Nemerow index method and the analytic hierarchy process are combined to comprehensively evaluate the two secondary indicators.

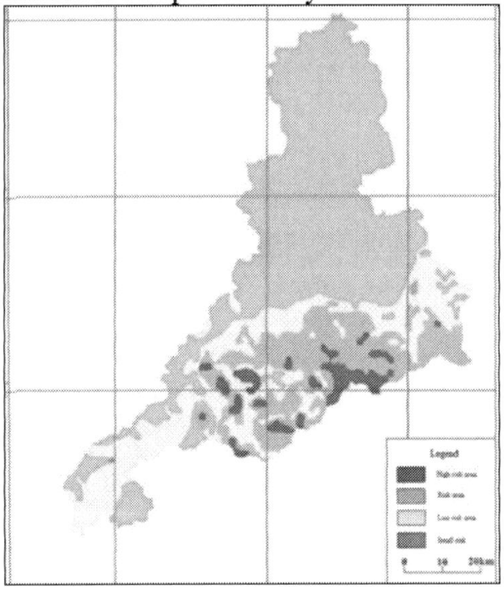

Fig. 2. Risk assessment results of primary environmental engineering geological problems.

 https://doi.org/10.1051/e3sconf/202234101015

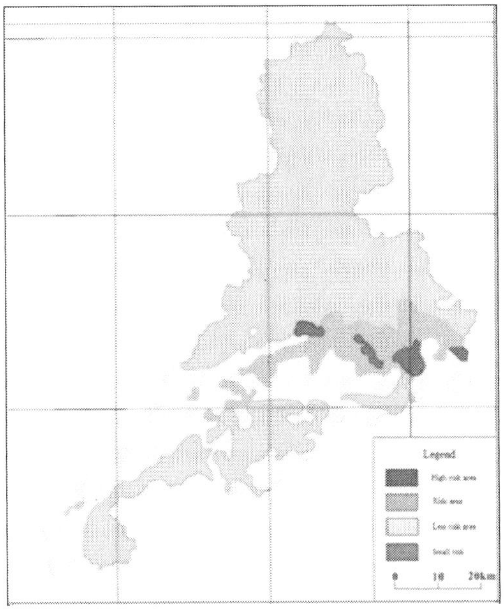

Fig. 3. Risk assessment results of engineering geological problems in secondary environment.

The Nemerow index method was used to calculate the normalized values of primary environmental engineering geological problems index and secondary environmental engineering geological problems index.The primary and secondary environmental engineering geological problems of each grid were calculated according to the formulas. Safety grade and description of major environmental engineering geological indexes are shown in Table 2.

Table 2. Safety grade and description of major environmental engineering geological indexes.

Security level	Level description
Class A [0, 2.5]	This level is the lowest safety level and is not suitable for major environmental engineering construction. Involving primary environmental factors and secondary environmental factors are inappropriate, human activities are strong, and population density is high.
Class B [2.5, 5]	This level is a restricted safety level. When major environmental projects are being constructed, it is more restricted by primary and secondary environmental factors, and human activities are relatively strong.
Class C [5, 7.5]	This level is a selective security level. During the construction of major environmental projects, it is less restricted by primary and secondary environmental factors, and the intensity of human activities is medium.
Class D [7.5, 10]	This level is the highest safety level and is suitable for major environmental engineering construction. Some areas are affected by a single factor, but the impact is relatively small.

$$P_{\mathrm{j}} = \sqrt{\frac{F_{\mathrm{jmax}}^2 + (\overline{\mathrm{F}}_{\mathrm{j}})^2}{2}} \quad (1)$$

Where: F_{jmax} is the largest single factor environmental quality index.

$\overline{F}_j$ is the average value of each single factor environmental quality index.

The evaluation criteria are shown in Table 2.

The evaluation of the engineering geological problems of the original environment in the grids with ID=1394 and 1395 can be obtained as follows:

$$X_{\text{native 1394}}=\frac{6.5248-5.0075}{8.6808-5.0075}\times 10=4.1305,\quad X_{\text{native 1395}}=\frac{6.2525-5.0075}{8.6808-5.0075}\times 10=3.3894$$

The evaluation of engineering geological problems of secondary environment in grids WITH ID=1394 and 1395 can be obtained.

$$X_{\text{secondary 1394}}=\frac{7.4598-4.1895}{8.7500-4.1895}\times 10=7.1708,\quad X_{\text{secondary 1395}}=\frac{6.2525-4.1895}{8.7500-4.1895}\times 10=7.8303$$

According to the grading standards, the security level of the grids with ID=1394 and ID=1395 is C. Other grids were graded according to the above evaluation method, and the vector diagram of the comprehensive evaluation zone was obtained, as shown in Figure 3.According to the above figure, the geological safety evaluation of major environmental engineering in Jinan is divided into four grades, namely Grade A, B, C and D.

Table 3. Natural-human environmental engineering geological index calculation table.

Cell	Area(km^2)	Perimeter (km)	Primary environment engineering geology index value	Engineering geological index of secondary environment	Nemerow index method calculated after the value
1394	4	8	4.1305	7.1708	6.4556
1395	4	8	3.3894	7.8303	6.8111

The area of Class A security zone is about 398.46km^2, accounting for 4.97% of the whole city.

The area of Class B security zone is about 1467.52km^2, accounting for 18.30% of the whole city.

The area of Class C security zone is about 1956.53km^2, accounting for 24.40% of the whole city.

The area of Grade D security zone is about 4194.80km^2, accounting for 52.32% of the whole city.

Fig. 4. Safety evaluation results of urban environmental engineering geological problems.

4 Conclusions

Urban environmental engineering geology is the core of environmental engineering geology research. At present, it mainly studies the regional adverse engineering geological effects in urban development, urban geological disasters, seismic, geological problems, geological problems caused by human engineering economic activities, urban water and soil pollution, and heavy metal pollution.The key problems to be solved in urban environmental engineering geology include establishing and improving the database of environmental quality of urban environmental engineering geology and predicting and evaluating the quality of environmental engineering geology in urban three-dimensional space

The nemerow index method is used to comprehensively evaluate the above two problems, and the safety zoning of environmental engineering geological problems in Jinan is obtained. It can be divided into four types: A, B, C and D. The area of grade A safety area is about 398.46km^2, accounting for about 4.97% of the city's area. The area of grade B safety area is about 1467.52km^2, accounting for 18.30% of the city's area. The area of grade C safety area is about 1956.53km^2, accounting for 24.40% of the city's area. The area of grade D safety area is about 4194.80km^2, accounting for about 52.32% of the city's area. Compared with the original evaluation results, this comprehensive evaluation method of "primary and secondary" environmental engineering geology is more scientific and more consistent with the actual situation.

References

1. X.C.Ling, L.X. Wang, H. Zhou, J. Some environmental engineering geology problems in urban construction,Harbin Univ. Eng. Architect. **33**,30(2000)
2. J.M. Lee, Study on urban engineering geological conditions and geological problems, Eng. Constr. **25**,4(2011)
3. C.E. Feng, J.W. Zhou, A.G. Zhou, Chin. J. Current development characteristic and tendency of environmental geology in the world and analysis of environmental geology problems in our country,Geol. Hazard Control **15**,9(2004)
4. C.S. Wang,W.M. Ye,H.J. Du, Chin. J. The Urban Geo-Environmental Effect of Underground Engineering in Shanghai City ,Undergr. Space Eng. **1**,283(2005)
5. G.H. Luo,C.H. Yan, X.Z. Lee,B.T. Xu, J. Study on urbanization and urban environment geotechnical engineering and geological hazard,Eng. Geol. **12**,125(2004)
6. Y.J. Chen,C.H. Zhang, Resour. Discussion on key techniques and problems of modern urban environmental engineering geology ,Economization Environ. Prot. **(3)**,51(2015)
7. X.R. Lee,S.H. Zhang, J. Geol. Study on systematic prevention and control of environmental engineering geological problems in coastal cities, Hazards Environ. Preservation **10**,20(1999)
8. L.P. Xing, J. Discussion on environmental engineering geology in urban construction,Henan Sci. Technol. **(7)**,174(2014)
9. R.G. Israfilov, Anthropogenic changes to hydrogeological conditions in urban areas,Urban Groundwater Manage. Sustainability **74**,11(2006)
10. G.R. Liu, Research on urban environmental engineering Geology .Hydrogeol. Eng. Geol. **4**,43(1989)

11. H.T. Liu, S.C. Gao, Analysis of environmental engineering geology problems in urban construction ,Forest. Sci. Technol. Inform. **41**,90(2009)
12. S.Y. Guan, S.D. Pang, Z.G. Long, Review on environmental and engineering geology prombles of Wuhan city,J. Eng. Geol. **15**,186(2007)
13. D.B. Wang, S.B. Yan, Discussion on urban environment engineering geology of Qingdao city,J. Eng. Geol. **14**,251(2006)
14. G.F. Lee, Problems of environmental engineering geology for city construction and sustainable development of Tangshan .J. Hebei Inst. Technol. **27**,121(2005)
15. X.L. Zhao, Exploration of environmental engineering geological problems ,Energy and Energy Conservation **(04)**,67(2013)
16. L.P. Xing, Study on the status quo of urban environment and urban engineering geology J. Henan Sci. Technol. **(14)**,174(2014)
17. Y.Q. Jia, Analysis of environmental engineering geology problems in urban construction ,Low Carbon World **(9)**,130(2015)
18. D.D. Zhang, Discussion on key techniques and problems of modern urban environmental engineering geology ,Tourism Overview **(11)**,233(2014)
19. S.J. Wang, Geological environment problems in urban development in China J. Quatern. Sci. **02**,115(1996)
20. L.J. Yue, Research status and existing problems of geotechnical geology of urban environment,China Homes,2011,(5),146.(2011)
21. Mejía-Navarro Mario, E. Wohl Ellen, Geological hazards vulnerability and risk assessment using GIS:model for Glenwood Spring, Colorado ,D. Oaks Sherry, Geomorph. **10**,331(1994)
22. M. Fedeski , J. Gwiliiani, Urban sustainabiiity in the presence of food and geo logical hazards: The development of a GIS-based vulnerability and risk assessment methodology ,Landsc. Urban Plan. **83**,50(2007)
23. F. Guzzetti, A. Carrara. M. Cardinali. P. Reichenbach, Landslide hazard evaluation: a review of currentte chniques and their application in amultiscale study. Central Italy ,Geomorphology **31**,181(1999)
24. M. d Ayman Hassaan Ahmed, E.S. Marwa Adel, Development of sustainable urban green areas in Egyptian new cities :The case of EISadat City, Urban Plan. **101**,157(2011)
25. E. Alparslan, F. Ince, B. Erkan, C. Aydöner, H. Özen, A. Dönertaş, S. Ergintav, F.S. Yağsan, A. Zateroğulları, I. Eroğlu, M. Değer, H. Elalmış, M. Özkan , A GIS model for settlement suitability regarding disaster mitigation, a cases tudy in Bolu Turkey ,Eng. Geol. 96,126(2008)
26. L.J. Zhang, Study on urban environment and urban engineering Geology,Sci. Technol. **(4)**,30 (2014)
27. H.W. Lee, The study on urban engineering geological environmental quality in planning, J.W.Xu, J. Eng. Geol. 14,204(2006)
28. J.L. Jia, Systematic engineering analysis method of urban environmental engineering geology and its application ,Chin. Acad. Geol. Sci. 1991.
29. D.J. Du, Research route and comprehensive evaluation of urban engineering geological environment, Hydrogeol. Engi. Geol. (5),3(1989)
30. Y.G. Jia, Study on the stability of Qingdao urban geological environment system ,H.J. Liu, Coast. Eng. **(03)**,28(1998)

31. Y.D. Lu, J. Zhang, M.S. Lee, GIS-based evaluation of urban environmental engineering geology by fuzzy logic: a case study of LuyangCounty, Shaanxi Province,J. Nat. Disaster. **14**,93(2005)
32. J.X. Wang, Analysis and evaluation of engineering geological environment quality of Olympic Park, Chin. Univ. Geosci. Beijing 2006.
33. Q. Gao, Study on the theory and method of urban Engineering Geological environmental quality evaluation, Manage. Technol. SME **(9)**,154(2014)
34. X.R. Lee, H.Q. Quan, Z.B. Cao, BP model of neural network theory and its application to urban environmental engineering geology, Environ. Monitor. Chin. **17**,15(2001)
35. X.R. Lee,J.Q Guo,X.G. Zong, Study on the theory and method of urban Engineering Geological environmental quality evaluation ,J. Ningxia Eng. Technol. **1**,35(2002)
36. S.Q. Han, Comprehensive evaluation of engineering geological environment in Shijiazhuang city based on BP neural network ,Hebei Univ. Technol. (2005)
37. Z. Miao, Study on evaluation of Urban geological environment quality in Xi 'an based on GIS, Chang`an Univ. (2007)
38. Y.C. Zhang, Engineering geological modeling and environmental quality assessment of the construction layer in Tongzhou District of Beijing ,Chin. Univ. Geosci.,Beijing (2009)
39. X. Huang, Research on engineering geological environment and construction suitability of Beijing Yanqing Planning New Town ,Chin. Univ. Geosci.,Beijing (2013).

Discussion on rapid evaluation method of large-scale land degradation in China

Yu Xing*, *Yingjie* Zhou, *Weiling* Yao, *Hao* Wang, *Li* Li, *and Jie* Wang

China Aero Geophysical Survey and Remote Sensing Center for Natural Resources, 100083 Beijing, China

Abstract. This paper discusses the effects of land degradation assessment based on the index of net primary productivity (NPP), rain use efficiency of vegetation (RUE) and relative rain use efficiency of vegetation(rRUE). The results show that the land degradation evaluation method based on the correlation coefficient of NPP with time and the annual average and annual maximum of rRUE can comprehensively reflect the status of land productivity. It plays a good role in indicating the effect of land degradation, and is more conducive to remote sensing quantitative evaluation and rapid monitoring of large-scale land degradation.

1 Forward

The traditional land degradation assessment method is to carry out field investigation or laboratory records of soil at a specific point in a specific area. This method is suitable for monitoring and evaluation of point level or small regional scale and less field investigation, but not suitable for large-scale remote sensing monitoring of land degradation. With the development of remote sensing and GIS technology, it is possible to realize mesoscale or large-scale remote sensing evaluation and monitoring of land degradation in time and space, and the idea of land degradation assessment and monitoring technology based on "3s" is becoming more and more clear. However, most of them obtain land type information through visual interpretation of remote sensing images, and then combine with ground survey to semi-quantitatively evaluate land degradation and its impact. Due to regional differences and multi-source data, there are a variety of land degradation indicators; classification standards are not consistent; standardized quantitative indicators are lacking; and most of them are indirect indicators, which are difficult to obtain through remote sensing technology. There are few studies on large-scale rapid quantitative assessment of land degradation. Accurately obtaining soil information and vegetation information by remote sensing, and monitoring the changes of various soil and vegetation indicators by remote sensing in the process of land degradation is the basis of remote sensing quantitative evaluation and monitoring. Using NPP and RUE indicators, scholars have achieved good results in monitoring the scope and degree of land degradation. For example: Yin He et al.[1] selected the RUE evaluation index of high-temporal resolution of remote sensing inversion and used the time series analysis method to evaluate the development trend of

*Corresponding author: 63216429@qq.com

desertification in Inner Mongolia Autonomous Region. Wang Hongyan[2] carried out remote sensing monitoring and evaluation of land degradation in the wind and sand source areas of Beijing and Tianjin based on NPP and RUE. Gao Zhihai et al.[3] discussed the remote sensing evaluation method for desertification based on RUE. In this paper, three indexes suitable for large-scale net primary productivity (NPP), rain use efficiency of vegetation(RUE), relative rain use efficiency of vegetation (rRUE) are compared and analyzed, and the optimal method for evaluating land degradation is obtained.

2 Comparison of evaluation methods

• Calculate the *RUE*, that is, the ratio of *NPP* to precipitation (*P*):

$$RUE=NPP/P \tag{1}$$

• The annual moisture index (MI) from 2000 to 2010 is calculated, and the climatic regions are divided at fixed intervals (0.15 in this paper) to form a climatic gradient zone (18 climatic zones are divided in this paper).

• According to the probability of the RUE value in each type of climatic zone divided by the MI, the RUE values corresponding to the 95% and 5% lower quantile of the RUE probability distribution are selected as the upper limit value RUE0.95 and the lower limit value RUE0.05, which can better express the maximum and minimum values of relative land production capacity in each climatic zone.

• According to the maximum value of RUE, the minimum value of RUE in each type of climate gradient zone, and the median value of MI interval in each type of climate gradient zone, the fitting function equation, namely the edge function model, is established.

• Taking the spatial raster data of MI as independent variables, the spatial raster data of the maximum and minimum values of RUE in China are obtained by spatial simulation according to the edge function model.

• The rRUE is calculated, and the threshold analysis is used to grade the land production capacity of rRUE. The *rRUE* calculation formula is

$$rRUE=(RUE-RUE_{0.05})/(RUE_{0.95}-RUE_{0.05}). \tag{2}$$

In the formula, *rRUE* represents the position of each pixel RUE within the range of the maximum and minimum potential RUE in the climate gradient, and between [0,1]; $RUE_{0.05}$ and $RUE_{0.95}$ represent the minimum and maximum values of potential rain use efficiency, respectively.

• The spatial distribution maps of rRUE in different periods are generated. The rRUE maps eliminates the influence of large-scale climate gradient. The RUE edge function is the natural "benchmark" of vegetation potential productivity in the climate region. Therefore, any land cover type in the region can be directly compared.

• Annual MI refers to the ratio of annual precipitation(*P*) to potential evapotranspiration(*PE*) in the same period [4].

3 Results and analysis

The spatial distribution maps of annual average NPP and annual RUEmean in China are calculated, as shown in Fig.1. The spatial distribution maps of rRUEmean and rRUEmax in China are calculated, as shown in Fig.2.

 https://doi.org/10.1051/e3sconf/202234101016

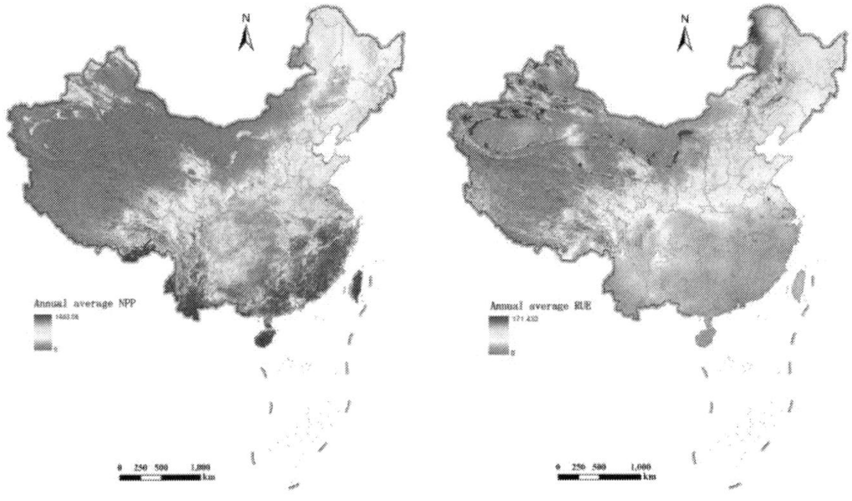

Fig. 1. Spatial distribution of annual average NPP (left) and annual RUE (right) in recent 10 years.

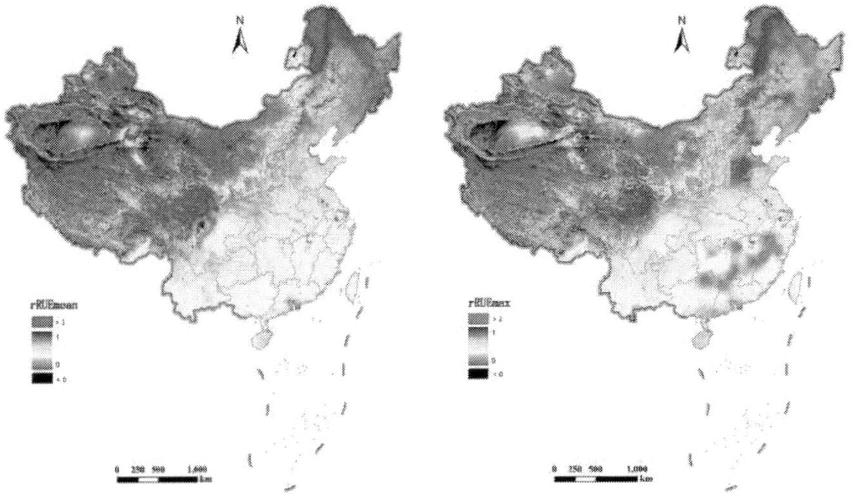

Fig. 2. Spatial distribution of rRUEmean (left) and rRUEmax (right) in recent 10 years.

From the analysis of the calculation results of NPP and RUE, the spatial distribution of RUE in China is very different from that of NPP. In southeast China, the annual NPP value is high, but because the precipitation in this region is also abundant, the RUE value is low. In northeast China, because the vegetation NPP is significantly smaller than that in the southeast, and because the difference in precipitation is smaller than that in NPP, the RUE value in the northeast is higher. In northwest China, there are some areas where the NPP value is low, but the RUE value is high, which is due to the small precipitation. Although the vegetation productivity shows an upward trend and the productivity has increased, it does not mean that the land condition in this area has improved, nor does it mean that the land has not been degraded. When the vegetation productivity shows a downward trend, it can not fully show that the land is degraded, perhaps due to the reduction of precipitation. The actual situation of the land has not become worse.

Although both NPP and RUE indicators play an indicative role in the evaluation of land degradation, through the test and analysis of the change trend correlation analysis methods of NPP and RUE, it can be found that the statistical data of the two indicators reflect different land degradation trends. According to the analysis of spatial distribution, the change trends of NPP and RUE have no obvious law, some changes are opposite, or some changes are consistent. Therefore, the two indicators of NPP and RUE in the evaluation of land degradation need to be improved and need to be used comprehensively.

Land degradation is relative to the initial or previous state, which involves the benchmarking of land degradation. The "benchmark" of land degradation assessment refers to the maximum potential state that the ecosystem can achieve under certain climatic conditions [5], that is the final stable state that can be achieved by local natural vegetation without human disturbance [6]. RUE is a comprehensive indicator of ecosystem vegetation productivity and health. NPP and RUE represent the status of vegetation productivity. They are two important indicators to measure land productivity and effectively represent the development trend of land degradation. However, RUE is essentially a ratio. Obviously, if the vegetation productivity of the 2 areas is the same, but their different climatic conditions, the RUE value in arid areas with less annual precipitation is higher. Therefore, for such a large range of China, there is an obvious climatic gradient, which cannot be directly compared with RUE. Instead, the climate impact needs to be removed and the index values of different climate gradient zones can be compared on the basis of "relative benchmark".

On the basis of considering the climate "relative benchmark", the significance test of the correlation coefficient of NPP with time and the annual average and annual maximum of rRUE are used to comprehensively reflect the status of land productivity, which has a good indicative effect on land degradation.

4 Conclusion

The land production capacity is evaluated by multi-year average rRUE and annual maximum rRUE, and the time series NPP change is used to monitor vegetation degradation. Finally, considering the results of land production capacity evaluation and vegetation degradation monitoring, the average status of land degradation in China is evaluated. The rRUE of this method can comprehensively reflect the status of land productivity, has a good indication of land degradation, can reflect the objective situation of land degradation, and is more conducive to remote sensing quantitative evaluation and rapid monitoring of large-scale land degradation.

Acknowledgements

This paper is supported by Programs of China Geological Survey (Remote sensing monitoring of national mining exploration and ecological spatial in key areas).

References

1. H. Yin, Z.G. Li, Y.L. Wang, F. Cai, Acta Geographica Sinica,**66**,653 (2011)
2. H.Y. Wang. Assessment and monitoring of land degradation using NPP and vegetation rain use efficiency,*Chinese Academy of Forestry* (2013)
3. Z.H. Gao, Z.Y. Li, G.D. Ding, L. Y. Li, Science of Soil and Water Conservation, **3**, 37 (2005)
4. S.W. Zhuang, H.C. Zuo, P.C. Ren, G. J. Xiong, B. D. Li, W. C. Dong, L. Y. Wang, Climatic and Environmental Research, **18**,617(2013)
5. P.P. Mi. The designing research on land degradation monitoring system, *Northeast Agricultural University*(2008)
6. Y.P. Liu, Journal of Arid Land Resources and Environment, **12**,74(1998)

Study on the construction of underground reservoirs on Zhoushan Island and related environmental issues based on the use of typhoon rain

Zhou Ye*, *Danna* Zhou, and *Fang* Chen

Zhejiang Ocean University, School of Marine Engineering Equipments, Zhoushan 316004, Zhejiang, China

Abstract. Water resources in the island area vary greatly from year to year and are unevenly distributed within the year, especially during typhoon and rainstorm periods when the lack of reservoirs leads to a large loss of rainfall and flood resources. In order to improve the water resources supply capacity of the island, this paper takes the excavated underground reservoir on Zhoushan Island as the research object to solve the problem of abundant rainfall and flood resources but lack of saving space during the plum and typhoon floods, and to enhance the self-sufficiency of water resources on the island. The study focuses on three environmental issues in the construction of underground reservoirs, and analyses the actual effect of the resourceful use of rainwater floods using the actual measurement data of Typhoon 2106 (In fa) to prove the feasibility of the construction of underground reservoirs. It makes a useful exploration for the future planning and construction of underground reservoirs on Zhoushan Island using underground space resources to solve water shortage.

1 Introduction

Many studies have been done by domestic scholars in the past. You Chunfeng[1] and others transformed the reservoir into a "combined surface-underground" reservoir through vertical impermeability treatment of the dam base, which improved the storage capacity and water supply assurance of the reservoir; Li Tong[2] aimed at maximising the utilization of water resources and minimising water abandonment after combined surface water and groundwater regulation, and established Yang Yun et al.[3] used operational research methods to establish an optimal dispatching model for the joint use of multiple reservoirs for water supply, simulating the joint use of plain reservoirs and underground reservoirs; Wang Xingchao[4] demonstrated the feasibility of underground reservoirs in sponge city construction through basic issues such as storage, purification and water circulation paths; Wang Congrong et al.[5] considered that pit-type underground reservoirs would be an important development direction in this field, and proposed that the water quality and quantity of the rechargeable water source should be considered comprehensively. This

* Corresponding author : 3043109291@qq.com

paper takes Zhoushan Island as an example to study open underground reservoirs and related environmental issues, and to explore the technology of resource utilisation of rain and wind on the island platform.

2 Zhoushan Island water system and underground reservoir scheme

2.1 Overview of the Zhoushan Island water system

The city's average annual precipitation is 1275.2mm, and the city's average maximum annual rainfall is 1648.5mm (1967). Zhoushan Island has no large water system, and the rivers on the island do not affect each other and flow into the sea alone. The upper reaches of the river basins are mountainous rivers, with high and low floods; the lower reaches enter the plains, where the river slopes gently and are influenced by the tides, with the characteristics of coastal plain rivers. The topographic map of the Zhoushan Island water system is shown in Figure 1.

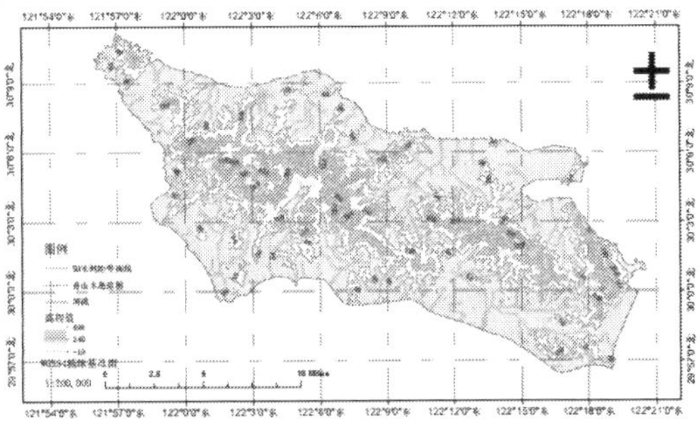

Fig. 1. Topographic map of Zhoushan Island water system

2.2 Causes of water shortage on Zhoushan Island

2.2.1 Difficulties in storing water in short and fast flowing rivers

The topographical conditions make it difficult to store water in the short flow of the source, and most of the rivers have already been "built as reservoirs and ponds", with limited potential for subsequent conventional surface water development. The total storage capacity of the island's reservoirs, ponds and rivers is about 110 million m^3. 370.00 km^2 of rainwater is collected in the mountains, and 150.00 km^2 of rainwater is collected in the reservoirs and ponds, with a utilization rate of more than 30%.

2.2.2 Uneven spatial and temporal distribution of rainfall

Although the average annual rainfall of Zhoushan Island is 1399.50 mm[6] and the evaporation of land surface is 758.30 mm, the total water resources are 300 million m3[7] , there is also an uneven distribution of rainfall between and within years. 1980-2020 annual precipitation ranges from 872 to 2108 mm, with an extreme ratio of 2.4 times. During the typhoon period from July to October, precipitation accounts for 35-45% of the annual precipitation. The monthly precipitation for 2018 and 2019 at Dinghai Station on Zhoushan Island is shown in Figure 2.

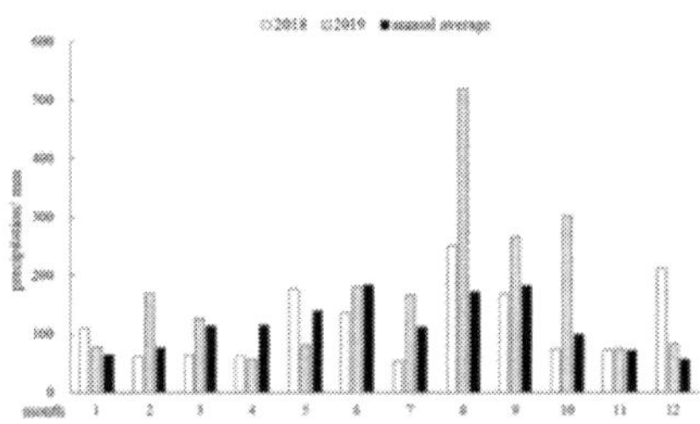

Fig. 2. Monthly precipitation at Dinghai Station, Zhoushan Island

2.2.3 Substandard water quality in mainland diversions

The first phase of the mainland water diversion project has a water supply scale of 1 m^3/s (86,000 m^3/d), with a total annual average diversion volume of 21.6 million m3. The second phase of the mainland water diversion project has a design diversion flow of 2.80 m^3/s, with an average annual diversion volume of 66.33 million m3. The third phase of the Zhoushan mainland water diversion project has a design diversion flow of 1.20 m^3/s, with an average diversion volume of 32 million m^3. However, there are many uncertainties such as interruptions due to damaged or burst water pipes; at the same time, the water quality is heavily polluted, with iron and manganese levels in the water bodies exceeding the Environmental Quality Standard for Surface Water (GB3838-2002).

2.2.4 Desalination is not yet widespread

According to China.org Ocean China 2019, Zhoushan has built 17 desalination plants and 26 desalination projects, with a desalination capacity of 137,000 tons per day. However, there are also problems such as high cost of water production, poor stability of water quality and low awareness of public acceptance.

2.3 Use of typhoon rain

2.3.1 Typhoon trends

Among the 46 typhoons that have landed in Zhejiang since the founding of the People's Republic, the number of typhoons that have landed in Zhejiang since 1978 is 33, i.e. 2.5 times that of the first 30 years in the last 40 years. 21 typhoons have landed between 2000 and 2021. Since the founding of the People's Republic of China, there have been 10 strong typhoons of magnitude 14 or above, 2 before 2000 and 8 after 2000. 75% of the typhoons that have landed in Zhoushan since 2000 have done so. Table 1 shows the impact of typhoons landing in Zhejiang on the rainfall at Dinghai Station on Zhoushan Island.

Table 1. Impact of typhoons registered in Zhejiang on rainfall at Dinghai Station, Zhoushan Island since 2000

Typhoon Landfall Range	Number of typhoons	100mm or more	50-100mm	Under 50mm
Zhejiang-Fujian Province border - Pingyang County	5	20%	20%	60%
Wenzhou City - Wenling City	8	50%	37.5%	12.5%
Jiaojiang City - Xiangshan County	4	75%	0%	25%
Zhoushan Island and Hangzhou Bay	4	75%	0	25%

2.3.2 Typhoon stormwater resourcefulness possibilities

There are 77 reservoirs on Zhoushan Island, with a normal storage capacity of 78.528 million m^3, a dead storage capacity of 950,000 m^3, a rainfall catchment area of 113.4km^2, a water diversion area of 10.73km^2, a multi-year average runoff of 74.484 million m^3 and an average storage capacity coefficient of 1.0415, but there is also an uneven inter- and intra-annual rainfall distribution. The storage capacity coefficient of Zhoushan Island reservoirs is significantly higher than that of mainland reservoirs, so the potential for dispatching is not great from this point of view, but from the changes in typhoons over the past 20 years, there is still great potential. However, judging from the trend of typhoon changes over the past 20 years, there is still a huge potential.

2.3.3 Underground Reservoir Programme

The special planning for water supply in Zhoushan City predicts that the per capita water consumption target for Zhoushan City in 2030 will be 450 L/day-per person, and the total water supply for Zhoushan Island in 2030 will be 82.125 million m^3. 30 million m^3 of underground reservoir capacity is proposed, and the total capacity of surface and underground reservoirs will be 60 million m^3, which will ensure water for production and living at a 90% guarantee rate. The water quality and cost of the reservoirs will be used as indicators, and the storage level of the underground reservoirs must not be lower than the average tide level for many years to prevent seawater from backing up and causing pollution to the water quality.

3 Environmental issues related to groundwater reservoirs

3.1 Ecological protection issues in the watershed

Due to the shortage of water resources, the reservoir capacity factor (reservoir capacity/total annual runoff) of the reservoirs built in Zhoushan has reached 1.2 on average, while that of other cities and counties in Zhejiang is roughly 0.3. Building underground reservoirs will further expand the reservoir capacity factor and aggravate ecological damage, and the ecosystem is weak, there is an urgent need to strengthen the ecological management of the compound basin. In the "Zhejiang Province Water Resources Department on strengthening the construction of beautiful rivers and lakes" and "Zhejiang Province Water Ecological Environmental Protection "14th Five-Year Plan" are proposed to put water ecological protection in an important position. Therefore, the reservoir area should be greened and the side slopes and ditches should be treated to control soil erosion. The original ecological environment downstream of the watershed should be considered to ensure the flow of ecological environment in each watershed after the completion of the reservoir and dynamically monitor the interaction effect between the underground reservoir and ecological.

3.2 Geological lithology issues

The distribution and development of the Quaternary sediments in the survey area are mainly controlled by geomorphology and sea level rise and fall, with complex genetic types, mainly Upper Pleistocene alluvial lacustrine (al-lQ3) clay, silty clay with silty clay and muddy sand and gravel, etc.; Holocene marine sedimentary (mQ4) silty clay, with residual slope deposits (el-dlQ4) accumulating on the slopes and along the foot of the slopes. The bedrock type of Zhoushan Island basically belongs to impermeable tuffs and fused tuffs, and the mountain is strong, with engineering geological conditions for the construction of underground reservoirs.

3.3 Groundwater reservoir water quality assurance issues

In terms of water quality testing, a study was carried out on the changes in water quality in the underground pit water storage in Dingjiashan, Mamu, Zhoushan City [8]. The results show that: the overall situation of water quality is good, but there are also problems such as the low number of indicators tested; limited space for experiments on water storage in the pit; and inconsistent water quality testing standards and methods. The test results are shown in Table 2.

Table 2. Test items in the Mamu Dingjia cave bank

Test unit	Unit 1		Unit 2		Unit 3	
Test time	17.4-12 (28 times)	18.1-7 (5 times)	18.4 - 19.1 (5 times)	19.4-9 (5 times)	19.3-8 (2 times)	19.4-9 (5 times)
Test indicators	PH, ammonia nitrogen, permanganate index and 8 other indicators	Same as left	PH, ammonia nitrogen, permanganate index and 7 other indicators	PH temperature, turbidity, dissolved oxygen	(GB5749-2006) Tables 1 and 2, (GB3838-2002) Tables 1 and 2 of 49 indicators	permanganate index and 9 other indicators

Thus, the following attempts can be made to safeguard water quality: i. routine 29 tests are carried out daily and 109 indicators are tested twice a year; ii. attention is paid to algae-derived odour substances such as 2-methylisobornyl (2-MIB), geosmin (GSM), β-violetone (β-Ion), β-cyclic citral (β-cyc), dimethyl trisulphide (DMTS), 2-isopropyl-3 methoxypyrazine (IPMP) and 2-isobutyl-3-methoxypyrazine (IBMP) and 2,4,6-trichloroanisole (TCA), which has an earthy smell; iii. Carry out regular dispatching to enable dynamic water flow in the reservoir and increase the amount of dissolved oxygen in the water.

4 Calculation of flood control dispatch during Typhoon 2106 (In fa)

4.1 Basic information on Typhoon 2106 (In-fa)

Typhoon No. 6 of 2021 "In fa" made landfall in Putuo District, Zhoushan at 12:30 on July 25, with a central pressure of 965 hPa and a maximum wind force of 13 near the centre (typhoon level), the fifth typhoon to make direct landfall in Zhoushan since 1949.From 8:00 p.m. on July 24 to 10:00 p.m. on July 28, 2021, the rainfall at the main 7 reservoir stations on Zhoushan Island was above 480mm, with a range of 481-536mm; the maximum hourly rainfall was 37mm, with a range of 23.5-37mm, and the maximum rainfall was at 1:00 p.m. on July 25, lasting for 1 hour. Rainfall duration was between 72-81 hours. 94 hours of rainfall of no more than 6 mm fell between 10.00 hours on 28 July and 8.00 hours on 1 August.

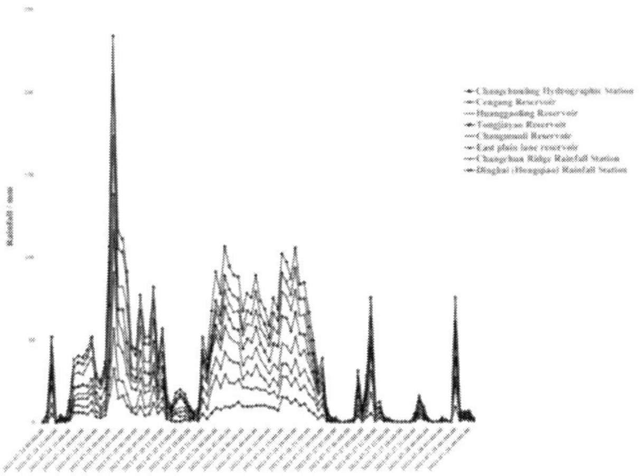

Fig. 3. Rainfall statistics at 8 measuring stations on Zhoushan Island during the typhoon 2106 (In fa)

4.2 Flood resource balance analysis during typhoons

In the calculation of the water balance of the operating reservoirs on this island, 77 reservoirs are calculated according to normal operation, the storage rate of the reservoirs that fail to operate normally is calculated according to the average storage rate of normal operation for each node, and the runoff coefficient is calculated according to 0.8, i.e. 400,000m³ of runoff volume is formed per square kilometre. After analysis, it can be obtained that the normal storage capacity of 77 operating reservoirs on the island during the typhoon is 79,483,000m³, the total area of rain collection is 123.69km², the formation of surface runoff is 49,476,000m³, the increase of water storage is 17,235,000m³, the pre-discharge is 1,526,000m³, the interception of water storage during the typhoon is 30,485,000m³, the total water discharge is 33,608,000m³.

Table 3. Summary of the water balance of the 77 operating reservoirs on the island

Region	Num ber of seats	Normal storage capacity	Rainwater catchment area	Total runoff	Previous storage capacity	Intermediate storage capacity	Later storage capacity	Impact overflow	Total discharg e flow
Dinghai District	52	3070.5	65.67	2626.8	1384.2	2899.7	2217	1111.3	1883.6
Water Company	7	3225	32.28	1291.2	2001	2922	2532	355.2	808.2
Putuo District	7	610.3	8.42	336.8	195.1	351.9	349.7	276.4	282.2
XinCheng District	11	1042.5	17.32	692.8	591.3	1042.5	815.02	241.6	476.6
Total	77	7948.3	123.69	4947.6	4167.7	7216.1	5891.1	1679.2	3360.8

4.3 Results of reservoir scheduling calculations during typhoons

Water resources dispatch calculation using the influence of typhoon 2106: 77 surface reservoirs and 30 million m³ of underground reservoirs on the island participate in the dispatch, and 49.476 million m³ of surface runoff is formed in the reservoir area, and the joint dispatch of surface and underground reservoirs can increase the water resources of underground reservoirs by 33.608 million m³. From the previous analysis, it can be seen that the average multi-year water resources distribution on Zhoushan Island is bimodal, i.e. 35-40% in the plum rainy season and about 35-45% in the typhoon rains. The rainfall of 513.5mm from the typhoon(In fa) is about 36.6% of the multi-year average rainfall of

1399.5mm, and a comprehensive consideration shows that the capacity of the chosen underground reservoir is suitable.

5 Conclusion

This paper studies the technical problems related to the construction of underground reservoirs on islands, and concludes that the excavation of underground reservoirs in island areas can make efficient use of rainwater resources, especially for the resourceful use of typhoon rain, and can effectively solve the shortage of water resources on islands. However, the construction of underground reservoirs will bring about many environmental problems. To ensure the safety of underground reservoirs and their water use, the study discusses four aspects, including typhoon change trends, ecological protection of composite watersheds, geological lithology and structure of the reservoir area and water quality assurance. This lays the foundation for subsequent research on the topic of underground reservoirs on Zhoushan Island. The specific site selection and rainfall resource scheduling mode of the underground reservoir on Zhoushan Island can be further analyzed, and the key technologies for the construction of the underground reservoir can be explored.

Acknowledgements

National Natural Science Foundation of China (51979249)

References

1. C. You, H. Zhang, The idea of converting Suihua Hongxing Reservoir into a “combined surface-underground reservoir”, Heilongjiang Science and Technology of Water Consercancy, **42(07)**, 238-240(2014)
2. T. Li, Joint regulation of surface water and groundwater considering flood resources, Dalian University of Technology, (2020)
3. Y. Yang, W. Wang, The joint use of plain reservoir and underground reservoir in Ganyu County, Jiangsu Province, China Rural Water and Hydropower, **01**, 26-30(2016)
4. X. Wang, Application of underground water reservoirs in construction of sponge cities, Advances in Science and Technology of Water Resources, **38(01)**, 83-87(2018)
5. C. Wang, A. You, L. Shu, Current situation and prospect on groundwater reservoir research, Zhejiang Hydrotechnics, **46(05)**, 68-71(2018)
6. T. Ho, H. Gong, An analysis of precipitation variation patterns in the past 64 years in the eastern Zhejiang islands — an example of Zhoushan Island,Zhejiang Hydrotechnics, **48(03)**, 20-21+33(2020)
7. X. Geng, Y. Jin, Research on building a security system of water resources in Zhoushan archipelago new are, Journal of Zhejiang Ocean University(Humanities science) , **32(01)**, 18-23(2015)
8. Y. Weng, J. Ke, M. Xu, Feasibility analysis of the construction of an underground reservoir on Zhoushan Island, Zhejiang Hydrotechnics, **47(06)**, 77-80(2019)

Analysis of wind-photovoltaic-thermal-pumped storage integration system considering carbon emission

Haofei Chen[1], *Huimin* Yan[2,*], *Feng* Wu[2], *Tianye* Yang[2], and *Yang* Shen[2]

[1]Economic Research Institute, State Grid Jiangsu Electric Power Co., Ltd., Nanjing 210008, China
[2]College of Energy and Electrical Engineering, Hohai University, Nanjing 211100, China

Abstract. This paper presents a new evaluation model of wind-Photovoltaic (PV)-thermal-pumped storage integration system considering carbon emission. The model is constructed on the basis of comprehensively considering the construction cost, operation cost, start-up and shutdown cost of the unit, wind abandonment rate, PV abandonment rate and carbon emission cost, and fully utilizing the power curtailments of wind and PV. It analyzes the electric energy benefit and low-carbon economy in three different operation modes including wind-PV-thermal-pumped storage, wind-PV-thermal and thermal under the same load. The effectiveness and rationality of the evaluation method are verified by a simulation example. Combined with the dynamic carbon price, the carbon economy of three different operation modes is analyzed. The analysis results show that wind-PV-thermal-pumped storage integration system has the best electric energy benefit and carbon economy, and with the increase of carbon price, it has more development prospects.

1 Introduction

To achieve the target of carbon peak and carbon neutrality, the low-carbon development of energy system is an inevitable trend. Renewable energy plus energy storage is an effective model for low carbonization and flexible development of energy system. In 2020, the total installed capacity of grid-connected wind and solar power generation in China was 119.87 million kW, accounting for 62.8% of the total installed capacity of new power generation in 2020[1]. With the rapid development of renewable energy, energy storage technology is especially important. Multi-energy complementary system can maximize the absorption of new energy, improve system reliability, reduce pollutant emissions. How to achieve the maximum power efficiency, and realize the low-carbon operation of energy system is an urgent topic to be studied.

In recent years, scholars at home and abroad have done a lot of research on capacity configuration, optimal scheduling and economy[2]-[7] of multi-energy complementary systems. Li et al. (2018) [2] established an economic analysis model considering the over interests of power system and energy storage system. Qi et al. (2020) [3] establishes an

* Corresponding author: hhuyhm@vip.sina.com

economic model of wind-PV-thermal-pumped storage integration system in which the maximum consumption of wind, photoelectricity and water was taken as the objective function. Zhang et al. (2015) [4] puts forward a new unit commitment model based on cost-benefit analysis, which takes into account both the economy and reliability of wind-pumped storage system operation. Li et al. (2009) [5] establishes a quantitative evaluation model of wind-pumped storage integration system considering peak and valley electricity prices. Xie et al. (2019) [6] establishes the economic model of micro-grid planning. Ye et al. (2018) [7] constructs an index framework suitable for evaluating wind-PV-water complementary characteristics. Xiao et al. (2020) [8] and Duan et al. (2015) [9] analyzes the energy saving and emission reduction benefits of renewable energy. However, many literatures do not consider the benefits of carbon emissions, or carry out post-assessment of carbon emissions.

In this paper, a new economic model considering carbon emission is constructed on the basis of considering various costs of wind-PV-thermal-pumped storage integration system. It analyzes the cost-benefit and low-carbon economy in three different operation modes including wind-PV-thermal-pumped storage mode, wind-PV-thermal mode and thermal mode under the condition of having the same load.

2 Economic model of wind-PV-thermal-pumped storage integration system considering carbon emission

2.1 Multi-energy complementary cost evaluation model

Three regional power grid models (wind-PV-thermal-pumped storage, wind-PV-thermal, thermal) are constructed. Taking the year as the calculation unit.

2.1.1 The wind-PV-thermal integration system operation (model II)

(a) on-grid energy

$$Q_{wpc} = Q_{wind}(1-\theta_{gw})(1-\delta_{wind}) + Q_{pv}(1-\theta_{gp})(1-\delta_{pv}) + Q_{coal}(1-\delta_{coal}) \quad (1)$$

where Q_{wind}, Q_{pv}, Q_{coal} are the electric energy production of wind farm, PV power station and coal-fired power plant respectively and θ_{gw}, θ_{gp} are the abandoned wind rate of wind farm and the abandoned photoelectricity rate of PV power station respectively. δ_{wind}, δ_{pv}, δ_{coal} are the station service power consumption rate of wind farm, PV power station and coal-fired power plant respectively.

(b) Operating cost

$$E_{wpc} = (I_{wind}r_{wind} + O_{wind}) + (I_{pv}r_{pv} + O_{pv}) + (I_{coal}r_{coal} + O_{coal} + Q_{coal}\beta_{coal}P_C) \quad (2)$$

where I_{wind}, I_{pv}, I_{coal} are the construction cost of wind farm, PV power station and coal-fired power plant respectively. r_{wind}, r_{pv}, r_{coal} are the annual depreciation factor of wind farm, PV power station and coal-fired power plant respectively. O_{wind}, O_{pv}, O_{coal} are the annual operation cost of wind farm, PV power station and coal-fired power plant respectively. β_{coal} is the coal consumption per unit power generation. P_c is the price of coal.

(c) Carbon emission cost

In order to limit and reduce pollutant emissions, the national carbon trading market was officially launched, granting a certain amount of carbon emission quota, and trading the excess part in the market, so as to promote the adjustment of energy structure.

$$E_{wpc-c} = Q_{coal} \cdot \beta_{coal} \cdot B_i \cdot \gamma_i - e_{wpc-c} \tag{3}$$

where B_i is the environmental value of pollutants. γ_i is the emission coefficient of pollutants. e_{wpc-c} is the cost of carbon emission quota.

(d) Electrical energy benefit considering carbon emission

$$A_{wpc} = C_{wind} Q_{wind}(1-\theta_{gw})(1-\delta_{wind}) + C_{pv} Q_{pv}(1-\theta_{gp})(1-\delta_{pv})$$
$$+ C_{coal} Q_{coal}(1-\delta_{coal}) - E_{wpc} - E_{wpc-c} \tag{4}$$

where C_{wind}, C_{pv}, C_{coal} are the on-grid price of wind farm, PV power station and coal-fired power plant respectively.

2.1.2 The wind-PV-thermal-pumped storage integration system operation (model I)

The integration system operation of wind-PV-thermal-pumped storage is to use abandoned wind and abandoned PV power to pump water in pumped storage power station, which making full use of wind energy and solar energy resources, thus reducing the power generation of thermal power plant, increasing the start-up and shutdown costs of thermal power plant, and realizing the multi-energy complementary effect.

$$Q_{wind} \cdot \theta_{gw} \cdot (1-\delta_{wind}) + Q_{pv} \cdot \theta_{gp} (1-\delta_{pv}) = Q_{in.p} \tag{5}$$

$$\Delta Q_{coal} (1-\delta_{coal}) = \eta_p Q_{in.p} \tag{6}$$

$$\Delta C_{on-off} = \Delta B_{on-off} \cdot P_c \tag{7}$$

where $Q_{in.p}$ is the annual power consumption of pumped storage power station when storing water. η_p is the energy conversion coefficient of pumped storage power station. ΔQ_{coal} is the reduced power generation of coal-fired power plant after the construction of pumped storage power station. ΔC_{on-off} is the annual new start-up and shutdown costs of coal-fired power plant after the construction of pumped storage power station. ΔB_{on-off} is the new fuel (standard coal) which is added to start and stop the coal-fired power plant after the construction of pumped storage power station.

(a) Operating cost

$$\Delta E = \left(I_p \cdot r_p + O_p + C_{in.p} \cdot Q_{in.p}\right) - \Delta Q_{coal} \cdot \beta_{coal} \cdot P_c + \Delta B_{on-off} \cdot P_c \tag{8}$$

$$E_{wpch} = E_{wpc} + \Delta E \tag{9}$$

where I_p is the construction cost of pumped storage power station. r_p is the annual depreciation coefficient of pumped storage power station. E_{wpc} is the annual operation cost of pumped storage power station. $C_{in.p}$ is the pumping electricity price of pumped storage power station.

(b) Carbon emission cost

$$\Delta E_c = \left(-\Delta Q_{coal} \cdot \beta_{coal} + \Delta B_{on\ off}\right) B_i \gamma_i \tag{10}$$

$$E_{wpch-c} = E_{wpc-c} + \Delta E_c \tag{11}$$

(c) Annual electric energy benefit after construction of pumped storage power station

$$\Delta A = [C_{wind} \cdot Q_{wind} \cdot \theta_{gw} \cdot (1-\delta_{wind}) + C_{pv} \cdot Q_{pv} \cdot \theta_{gp} \cdot (1-\delta_{pv}) - C_{coal} \cdot \Delta Q_{coal} \cdot (1-\delta_{coal}) + C_{out.p} \cdot Q_{out.p} - C_{in.p} \cdot Q_{in.p}] - \Delta E - \Delta E_C \tag{12}$$

$$A_{wpch} = A_{wpc} + \Delta A \tag{13}$$

where $C_{in.p}$ is the pumping electricity price of pumped storage power station. $C_{out.p}$ is the on-grid price of pumped storage power station.

2.1.3 The thermal power plant operation (model III)

When a thermal power unit operates independently, its generating capacity should meet the demand of the total load, that is, the generating capacity of the thermal power unit should be equal to the total load electricity.

$$Q_c = Q_f = Q_{wpc} \tag{14}$$

(a) Thermal power unit capacity

$$M_c = Q_{wpc} / (1-\delta_{coal}) T_{coal} \tag{15}$$

where T_{coal} is the annual utilization hours of thermal power. Q_f is the total capacity of loads.

(b) Operating cost

$$E_c = I_{coal} r_{coal} + O_{coal} + Q_{coal} \beta_{coal} P_c \tag{16}$$

(c) Carbon emission cost

$$E_{c-c} = Q_{coal} \beta_{coal} B_i \gamma_i - e_{c-c} \tag{17}$$

where e_{c-c} is the cost of carbon emission quota.

(d) Electrical energy benefit considering carbon emission

$$A_c = C_{coal} Q_{coal} (1-\delta_{coal}) - E_c - E_{c-c} \tag{18}$$

2.2 Evaluation of carbon economy

2.2.1 Low carbon contribution rate

In order to better evaluate the carbon economy of multi-energy complementary system, low-carbon contribution rate was calculated. u_{c1} is the low-carbon contribution rate of model I relative to model III. u_{c2} is the low carbon contribution rate of model II relative to model III. The higher the value, the better the carbon reduction benefit.

$$u_{c1} = \frac{E_{c-c} - E_{wpch-c}}{E_{c-c} - E_{wpch-c} + C_{coal} Q_{coal} (1 - \delta_{coal})} \tag{19}$$

$$u_{c2} = \frac{E_{c-c} - E_{wpc-c}}{E_{c-c} - E_{wpc-c} + C_{coal} Q_{coal} (1 - \delta_{coal})} \tag{20}$$

2.2.2 Carbon emission compensation time

Based on the concept of investment payback period in engineering economics, the carbon emission compensation time is put forward to evaluate the years required for the carbon emission cost to compensate the annual construction cost.

$$T_c = \frac{E_c}{Q_{coal} \cdot \beta_{coal} \cdot B_i \cdot \gamma_i} \tag{21}$$

3 Analysis of simulation

3.1 Multi-Energy Complementary System Structure

In this paper, a region in northern China is selected as the research background, and the multi-energy complementary system consists of 300MW wind farm, 200MW PV power station, 1000MW thermal power station and 200MW pumped storage power station.

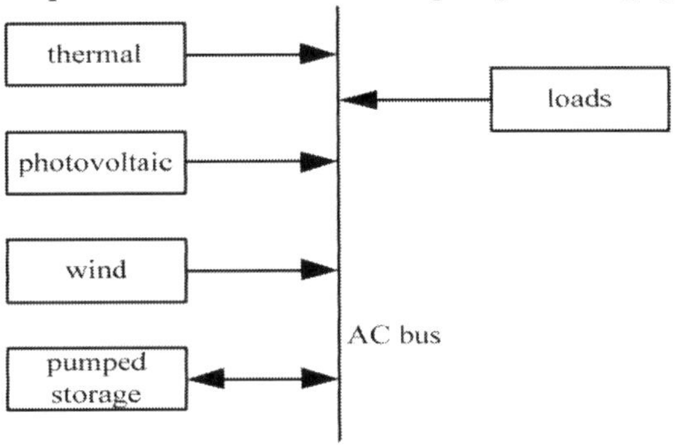

Fig. 1. Wind-PV-thermal-pumped storage integration system

3.2 Simulation Parameters

According to the data recommended by the National Development and Reform Commission, γ_i the emission coefficient of pollutants is 2.62, B_i the environmental value of pollutants is 0.2 Yuan/kg, β_{coal} the coal consumption per unit power generation is 0.32 kg/kWh[9], P_c the price of coal is 0.7 Yuan/kg, $\Delta B_{on\text{-}off}$ the new fuel (standard coal) which is added to start and stop the coal-fired power plant after the construction of pumped storage power station is 195 yuan/t and η_p the energy conversion coefficient of pumped storage power station is 0.75.

Table 1. System Parameter

Unit Type	Construction Cost(Yuan/kW)	Annual Operation Cost(Yuan/kW)	Station Service Power Consumption (%)	Abandoned Wind and PV Rate(%)	On-grid Price (Yuan/kW)	Annual Depreciation Factor (%)	Annual Utilization Time(h)
wind farm	6500	30	3.0	20	0.61	8	2000
PV station	5400	20	2.0	20	0.65	8	1300
pumped storage	4500	20	/	/	0.85(on grid) 0.21(pumping)	7	1000
thermal plant	3500	20	2.0	/	0.391	7	3900

3.3 Result Analysis

According to the calculation results and Figure 2, Figure 3, Figure 4 and Figure5: In the case of considering carbon emission cost, the annual operating cost from high to low is in the order of Model III > Model I > Model II; the carbon emission cost from high to low is in the order of Model III > Model II > Model I; the electric energy benefit from high to low is in the order of Model I > Model II > Model III. With the increase of carbon price, the carbon emission cost of the three operation modes increases gradually, the electric energy benefit decreases gradually, the low carbon contribution rate rises, the low carbon contribution rate u_{c1} is higher than the low carbon contribution rate u_{c2}, the carbon emission compensation time decreases, and model I shows a faster decline speed than model II and model III, which shows that model I has better economy and more development prospects.

Table 2. Calculation Results (I wind-PV-thermal-pumped storage system, II wind-PV-thermal system, III thermal system)

Model	Operation Cost (Hundred Million Yuan)	Unit Operation Cost (Yuan/kWh)	Carbon Emission Cost (Hundred Million Yuan)	Unit Carbon Emission Cost (Yuan/kWh)	Electrical Energy Benefit (Hundred Million Yuan)
I	15.0760	0.3430	0.9279	0.0211	4.0190
II	13.7400	0.3126	1.1038	0.0251	3.9320
III	15.6400	0.3559	1.1634	0.0265	-0.2850

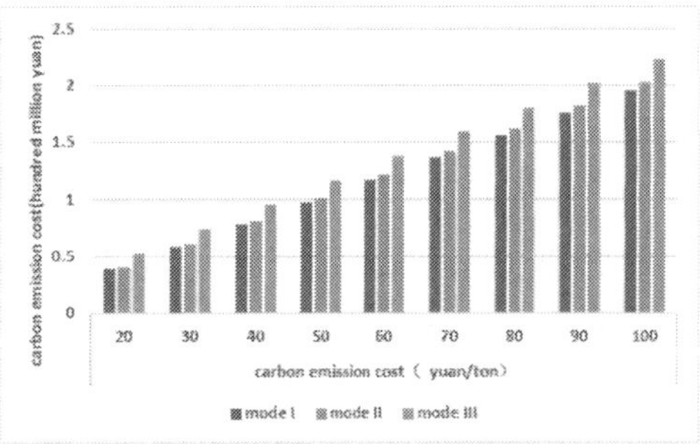

Fig. 2. Impact of carbon price on carbon emission cost

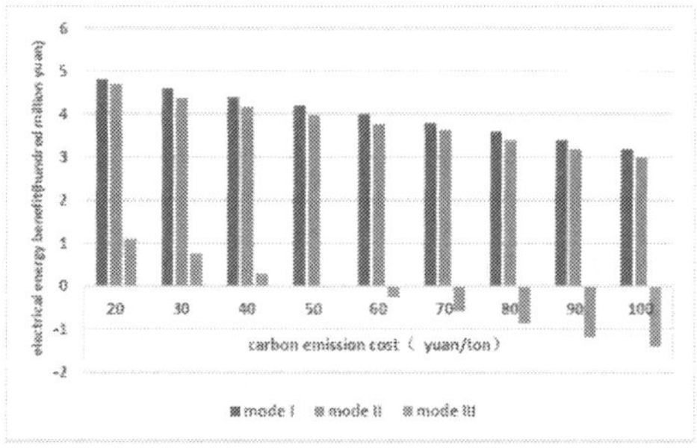

Fig. 3. Impact of carbon price on electrical energy benefit

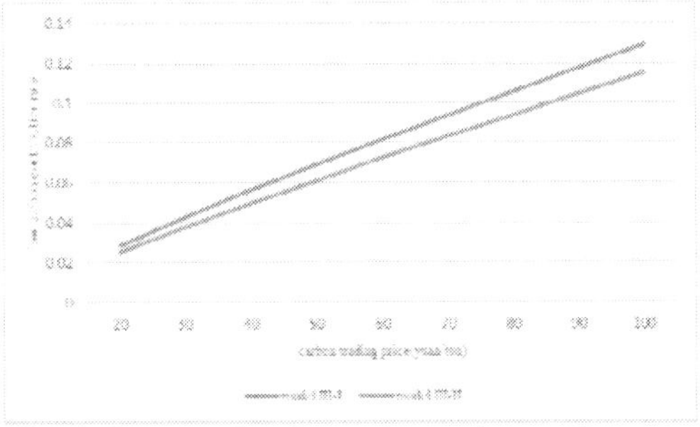

Fig. 4. Impact of carbon price on low carbon contribution rate

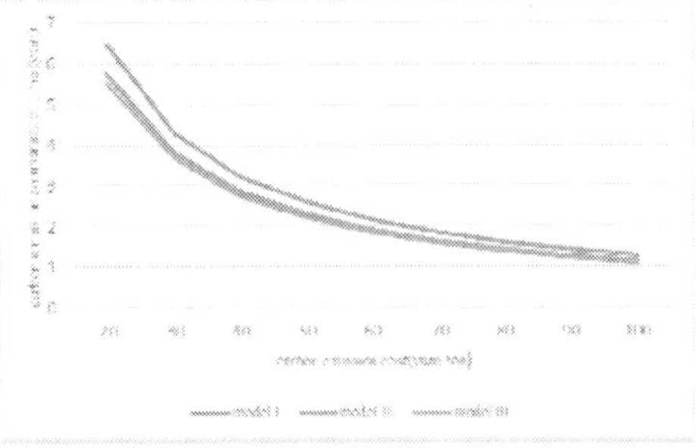

Fig. 5. Impact of carbon price on carbon emission compensation time

4 Conclusion

(1) After adding wind power and PV on thermal power plant, the operation cost is reduced by 19.0 million Yuan, the carbon emission cost is reduced by 1.2163 million Yuan, and the electric energy benefit is increased by 42.17 million Yuan. After adding wind power, PV and pumped storage on thermal power plant, the operation cost is reduced by 5.64 million Yuan, the carbon emission cost is reduced by 1.3922 million Yuan, and the electric energy benefit is improved by 43.04 million Yuan. The results show that considering the carbon emission cost, the thermal power plant has the highest carbon emission cost, the highest operating cost and negative electric energy benefit. Wind-PV-thermal-pumped storage system has the lowest carbon emission cost, and its operation cost is higher than that of wind-PV-thermal system, but its electric energy benefit is improved, and it has good economy.

(2) With the increase of carbon price, the cost of carbon emission is gradually increasing, and the carbon emission cost of thermal power is increasing rapidly. The electric energy benefit is gradually decreasing, especially in thermal power plants, where the loss is more serious, and the wind-PV-thermal-pumped storage system has the best electric energy benefit. With the increase of carbon price, the low-carbon contribution rate factor increases, and the low-carbon contribution rate u_{c1} is higher than the low-carbon contribution rate u_{c2}, the carbon emission compensation time decreases, and the model I shows a rapid decline rate, which shows that the model I (wind-PV-thermal-pumped storage system) has better low carbon economy and development prospects.

(3) In this paper, carbon emissions are included in the cost calculation of electric energy efficiency, which compresses the profits of enterprises. However, with the continuous decline of the cost of new energy, the proportion of new energy in the total energy is increasing, and the profits of enterprises will continue to rise. The model quantifies the power efficiency and carbon economy, which is conducive to guiding the planning and construction of new energy and providing theoretical basis for the system emission reduction scheme.

(4) Peak carbon and carbon neutrality is an energy revolution. At present, up to 51% of China's carbon emissions come from power generation and heat. Therefore, the energy structure should be adjusted. The launch of carbon trading market can change the energy structure from the supply side and guide users to use low-carbon energy through the price of electricity. In the future, thermal power enterprises are expected to transmit part of the carbon emission cost to the terminal, which will be borne by consumers of electric energy. As a major supplier of carbon dioxide emission reduction such as wind power and photovoltaic, it will definitely increase its income and become the main beneficiary. Therefore, after the operation of the carbon trading market, the carbon price will be coupled

with the cost of power generation, which will improve the electricity efficiency and promote the transformation of China's energy structure.

Acknowledgements

This work was supported by National Natural Science Foundation of China (51976053) and Program of Introducing Talents of Discipline to Universities (B14022).

References

1. X. H. Wang. China Energy Big Data Report (2021)-Electric Power. Electric Power Decision and Public Opinion reference [R], 2021,**500** (19).
2. J. H. Li, X. K. Fan, G. Mu, W. C. Ge, Y. F. Ge, D. C. Hu. Economic Analysis Of Energy Storage Applied To Grid Frequency Regulation[J]. Journal Of Global Energy Interconnection, 2018, **1**(3): 355-360.
3. Y. Z. Qi, Y. H. Huang, W. S. Wang, Y. F. Wang, X. Y. Pang, C. Zhang, H. T. Zhang. A Study on Hydro-Wind-Solar Consumption Analysis Method For High Proportion of Clean Energy[J]. Power System and Clean Energy, 2020, **36**(1): 55-63.
4. L. D. Zhang, M. H. Yin, J. Bu, K. L. Song, Y. Zhou. A Joint Optimal Operation Model of Wind Farms and Pumped Storage Units Based On Cost-Benefit Analysis[J]. Power System Technology, 2015, **39**(12): 3386-3392.
5. Q. Li, Y. Yuan, Z. J. Li, W. S. Wang, H. Y. Lu. Research On Energy Shifting Benefits of Hybrid Wind Power and Pumped Hydro Storage System Considering Peak-Valley Electricity Price[J]. Power System Technology, 2009, **33**(6): 14-18.
6. H. Xie, X. F. Teng, Y. J. Zhang, T. W. Zhen, L. J. Chen. Analysis Of Economic Influence Factors in Wind-PV-Storage Microgrid[J]. Automation Of Electric Power Systems, 2019, **43**(6): 70-76.
7. L. Ye, X. X. Qu, Y. X. Yao, J. T. Zhang, Y. F. Wang, Y. H. Huang, W. S. Wang. Analysis on Intraday Operation Characteristics of Hybrid Wind-Solar-Hydro Power Generation System [J]. Automation Of Electric Power Systems, 2018, **42**(4): 158-164.
8. J. W. Xiao, H. X. Huang, X. D. Liu. Carbon Economic Evaluation of Large-scale Wind-Photovoltaic-Storage Integration System[J]. Journal of Solar Energy, 2020, **41**(6) :171-176.
9. J. M. Duan, Z. X. Wang, C. M. Wang, Q. L. Zhou, J. H. Huan. Renewable Power Planning Considering Carbon Emission Reduction Benefits[J]. Power System Technology, 2015, **39**(1): 11-15.

Improved hydropower utilization design of the entire Wuding river in Yulin, Shaanxi

Tuo Liu*

Shaanxi Province Institute of Water Resources and Electric Power Investigation and Design, Xi'an, Shaanxi, 710001

Abstract. Below Cuijiawan, the main stream of Wuding River is basically a canyon section with exposed bedrock in the channel, intermittent plunging rapids, steep riverbed ratio and steep rock walls on both sides, which has favorable conditions for the construction of hydropower projects and is a key section for hydropower exploitation. Through research and analysis of the current water energy development and utilization, this paper combines the comprehensive improvement planning objectives and design ideas to propose a water energy utilization design that is consistent with the actual situation of Wuding River and reasonably budgeted, so that the abundant water energy resources of Wuding River can be fully utilized and the social, economic and ecological benefits can be unified.
Keyword: Wuding River; Water energy utilization; Generating capacity; Installed capacity.

1 Introduction

The Wuding River is a primary tributary of the Yellow River and the mother river of the Yulin people. It originates from the northern foot of Changchunliang, Baiyu Mountain in Yulin City, flows through Ordos City in Inner Mongolia, Yulin and Yan'an cities in Shaanxi Province, and joins the Yellow River at Hekou Village in Qingjian County. The Wuding River is 491.2 km long, with a basin area of 30,261 km^2. The Wuding River in Yulin is 442.8 km long, The watershed area is 20,302 km^2 [1].

Before the Qin and Han Dynasties, the Wuding River was known as the Sheyan River, and during the Northern and Southern Dynasties, it was known as the Xiashui and Shuofangshui. Since the Tang and Five Dynasties, the river has been called the Huanghudu River and the Wuding River because of its indefinite course, uncertain depth, indefinite volume, and indefinite clarity. Departing from the mountain stream area, it flows through the desert arid area all the way northward and passes through the southern edge of the Maowusu Sandy Land at the junction of Shaanxi and Inner Mongolia, turns eastward to Batuwan in Inner Mongolia, crosses the loess hills and ravines to the Yuhe Fort in Yulin and turns to flow southeast. Meandering southward, the river flows in the shape of a zigzag across the land of Yulin, like a kind mother with her warm arms open. The river nourishes 61% of the city's population, covers more than 80% of grain production regions, and contributes 42% of the city's total GDP.

* Corresponding author: 270689085@qq.com

The Wuding River basin occupies a strategically important position as a gathering place for urban population, a centralized zone for industrial development, a backbone hub for transportation, a benchmark for ecological balance, a water supply carrying area, and a core cultural landscape area in Yulin City. With the rapid development of the city's economy and society, the water security of the Wuding River also presents a severe situation where old and new problems are intertwined. The water quantity is decreasing rapidly, water quality is deteriorating, flood control capacity is decreasing, soil erosion is serious, and shoreline management is not standardized, which seriously affects the life of the river, the ecology of the river basin, people's welfare and economic development. Therefore, the Wuding River Basin is in urgent need of comprehensive improvement.

2 Development and utilization status

The main stream of Wuding River in Yulin City has a length of 442.8km, a drop of 589m, and an average annual flow of 36.5m^3/s. Theoretical hydro energy reserves in the basin are 209.0MW, including 163.8MW in the main stream, 43.3MW in installed capacity for technical exploitation, 43.30MW in installed capacity for economic exploitation, and 15.75MW in installed capacity for the exploited and under exploitation[3].

Below Cuijiawan, the main stream of Wuding River is basically a canyon section, with exposed bedrock in the channel, intermittent plunging rapids, steeper riverbed drop of more than 170m, a narrow valley with a width of 80m~120m, and steep rock walls on both sides, which has favorable conditions for the construction of hydropower projects and is a key section for hydropower development.

There is a hydrological station at Baijiachuan in this section of the river, with a control basin area of 29,662km^2, and the information from this station was adopted in this design scheme. According to the hydrological data, the multi-year average runoff of Baijiachuan hydrological station is 862 million m^3 with an average flow rate of 27.33 m^3/s. According to the analysis of the measured sediment data of Baijiachuan (Chuankou) hydrological station of Wuding River, the multi-year average sediment transport of suspended mass is 212 million t, and the annual average sediment transport of nudged mass is 42 million t. The multi-year average sediment transport of Baijiachuan hydrological station is 254 million t.

This section of the river is located in the composite part of the horseshoe-shaped betwixtoland of the zigzag Heilan Mountains and the northward slope of the northern Shaanxi in the Yikezhao League-Shaanxi Betwixtoland of the third neocathaysian subsidence zone. The bedrock is Paleozoic and Mesozoic sedimentary rocks, with a lot of Mesozoic rocks exposed, and a set of well cemented sandstones in the strata of dyas and triassic systems. The exposed strata in the area are Jurassic sandstone, and the attitude is nearly horizontal. Folds and faults are rare in this area. Since the Quaternary period, the crust in this area is mainly seismic uplift, with strong erosion and undercutting, and the development of gullies and valleys, forming the topography of loess hills and gullies. According to the Seismic Ground Motion Parameters Zonation Map of China (GB18306-2015), the basic intensity of the earthquake in this area is Grade VI.

Up to now, four hydropower stations have been built in the main stream of Wuding River from upstream to downstream in Yulin City: Wanggedu Reservoir Dam-toe Hydropower Station, Xiangshui Hydropower Station, Suide Hydropower Station and Dongfeng Hydropower Station. Except Wanggedu Reservoir Dam-toe Hydropower Station which is a reservoir hydropower station, the other three are diversion-type hydropower stations, all of which were built in the 1960s and 1970s and are still in normal operation. Wanggedu Reservoir Dam-toe Hydropower Station is located on the right bank of the dam, with a maximum diversion flow of 13.0 m^3/s in power generation and a minimum diversion flow of 8.82 m^3/s. The hydropower station has a guaranteed output of 1,575 kW, a total installed capacity of 3,750 (3×1,250) kW, and an average annual generating capacity of 19.4 million kW·h. Dongfeng Hydropower Station has an installed capacity of 6,000 kW, and an average annual generating capacity of 32 million kW·h. It is the largest hydropower station in Yulin City. The main indexes of hydropower stations available in the main stream of Wuding River are shown in Table 1.

Table 1. Main indexes of hydropower stations available in the main stream of Wuding River in Yulin City[4].

Item	Unit	Wanggedu Reservoir Dam-toe Hydropower Station	Xiangshui Hydropower Station	Suide Hydropower Station	Dongfeng Hydropower Station
Construction site	County (District)	Hengshan District	Hengshan District	Suide County	Qingjian County
Multi-year average flow rate	m^3/s	10.75	20	47	48
Development method		Dam-toe type	Diversion type	Diversion type	Diversion type
Normal water storage level	m	1046	970	771	618
Regulating performance		Years	No	No	No
Available drop (max/min)	m	8.1	31/30	12.9/13.5	29.8
Installed capacity	kw	3750	3600	2400	6000
Guaranteed output	kw	1575			
Annual generating capacity	10,000 kw·h	1940	1600	1320	3200
Power generation and diversion flow	m^3/s	13	15.0	26.2	24.0
Dam type		Homogeneous earth dam		Overflow dam	Overflow dam
Maximum dam height	m	44		7.3	7
Length of diversion channel	m	341.6		475	119

3 Design ideas

1) Adhere to the harmony of human and water, scientific development. Properly handle the relationship between small hydropower development and ecological protection of rivers, to achieve the greening of the whole process of small hydropower planning, construction, operation and management stages.

2) Adhere to land-specific conditions and promote by category. Taking full account of the actual situation in each place, the new small hydropower station should be prevented strictly from the source, without new debts. Small hydropower stations that have been built should urgently manage the ecological and environmental problems that exist, and repay old debts as much as possible.

3) Adhere to both incentives and constraints, with double-handed efforts. Formulate policies and measures to facilitate the construction of green small hydropower stations, while strengthening the strict control of the process, strict punishment for damage, and

accountability.

4) Adhere to the government-led approach and multi-party engagement. The government should develop green standards, and improve regulations and policies. Hydropower station owners should fulfill the main responsibility of its implementation, and play the role as social organizations and allow for the supervision of the public.

In terms of hydropower resources, we should adhere to the principles of moderate development, comprehensive utilization, balanced development and protection, correctly handle the relationship between hydropower development and ecological protection, and adhere to the priority of ecology:

1] Do not plan hydropower stations in all the rivers in the nature reserves, scenic spots and areas designated by law and defined areas prohibited from development, as well as sensitive areas of important ecological environment for endangered, rare and endemic protected animals and plants.

2] Prioritize the development of hydropower stations attached to the comprehensive use of water conservancy hub projects and hydropower stations by changing the function of the original water conservancy project or by tapping the potential of such a project.

3] Adjust the development methods of rivers to promote the development of rivers for the originally planned power stations with large losses due to migration and inundation.

4] Prioritize the planning of the leading cascade hydropower station to improve the efficiency of water resources utilization and the overall effectiveness of the basin development. Fully demonstrate the relationships between the cascades of the rivers mainly developed by diversion-type hydropower station, in order to reduce the length of the off-stream.

5] Prioritize the planning of hydropower stations with tourism and ecological functions, and realize more social functions of hydropower stations, so as to drive the rapid development of regional economy and society.

6] Consider the hydropower development and ecological protection of the main streams and tributaries, upstream and downstream, and make the organic connection between the development and planning of the main streams and tributaries.

4 Hydropower design

In the northern Shaanxi region, we should focus on ecological environment management and protection, and develop hydropower according to local conditions by combining the construction of photovoltaic and wind power projects in northern Shaanxi so as to realize the complementarity of hydropower, photovoltaic and wind power. At the same time, the technical transformation of the built hydropower stations should be carried out to increase the effective output. Dam-toe hydropower stations should be built in conjunction with the comprehensive utilization of water resources, in order to serve the water consumption of energy and chemical bases and urban and rural residents, and to improve the utilization rate of hydropower resources.

According to the "Shaanxi Province Small Hydropower Development and Utilization Plan" (revised) completed in 2016, the cascade sequence for the eight planned hydropower stations in the main stream of the Wuding River in Yulin City are: Wanggedu Reservoir Dam-toe Hydropower Station (established), Xiangshui Hydropower Station (established), Suide Hydropower Station (established), Lijiawa Hydropower Station, Cuijiata Hydropower Station, Chuankou Hydropower Station, Wangsuli Hydropower Station and Dongfeng Hydropower Station (established). These hydropower stations have a total installed capacity of 42.7MW, with an annual generating capacity of 183.49 million kW·h. As the planned Wangsuli Hydropower Station is located in the backwater area of Guxian Reservoir, the former level-4 stations of Lijiawa and Wangsuli is now adjusted to level-3 stations of Lijiawa, Dongfanghong and Xinchuankou. Accordingly, the design includes three new hydropower stations with an additional installed capacity of 31.05 MW and an annual generating capacity of 119.63 million kW·h. The main indicators of the proposed stations are shown in Table 2.

Table 2. Main indicators of the planned hydropower stations in the main stream of the Wuding River in Yulin City[2].

No.	Station name	County (district)	Township (town)	Development method	Maximum water head (m)	Installed capacity (MW)	Annual generating capacity (10,000 kW·h)
1	Lijiawa	Qingjian County	Dianzegou Township	Diversion type	31	7.65	3109
2	Dongfanghong	Qingjian County	Gaojiecun Town	Diversion type	43.1	15	5458
3	Xinchuankou	Qingjian County	Gaojiecun Town	Diversion type	33	8.4	3396
Total						31.05	11963

5 Conclusions

1) Developing hydropower resources with the principle of prioritizing ecology and building green small hydropower stations as the basic requirement.

2) Taking hydropower for people's livelihood, safe hydropower, green hydropower, and harmonious hydropower as the ultimate goal of hydropower resources development and utilization. We should initially form a new pattern of hydropower resources development, utilization and construction with the construction of water ecological civilization as the main line. Moreover, we should build a water security system that is compatible with the well-off society and promote the sustainable economic and social development of Yulin City.

3) Through rational planning, scientific planning and orderly development of hydropower resources, we should minimize the negative impact of hydropower projects on the ecological environment and keep the structure and function of river ecosystems in a sound state, so as to realize the unity of social, economic and ecological benefits.

References

1. The Bureau of Water Affairs of Yulin City, Shaanxi Province Institute of Water Resources and Electric Power Investigation and Design. Comprehensive Improvement Plan for the Entire Wuding River in Yulin City, 2017.
2. Shaanxi Provincial Water Resources Department. Planning for the Development and Utilization of Small Hydropower Projects in Shaanxi Province, 2016.
3. Yellow River Conservancy Commission of the Ministry of Water Resources. Comprehensive Plan for the Wuding River Basin, 2014.
4. Shaanxi Provincial Water Resources Department.Report on Integrated Planning of Wuding River Basin in Shaan xi Province,2012.

A two-stage inexact programming with value-at-risk for water resources management

Xiangming Kong*, *Donglin* Wang, *Chunxiao* Wang, *Yu* Wang, *Xiufeng* Yao, and *Xiaoming* Liu

Beijing Polytechnic, Beijing 100176, China

Abstract. In this study, the application of a two-stage inexact programming with value-at-risk (TIPV) model in water resources system planning has been developed. The TIPV method is intended to tackle the inexact parameters and the risks of economic loss. The application of case study shows that more alternatives under multiple levels of risks could be generated. The amount of water shortages and the width of system benefit intervals would decrease as the risk increases. TIPV could provide more effective information for stakeholders to recognize social policies with maximized system benefits under various risk levels.

1 Introduction

In water-resources system, many components and parameters are influenced by uncertainties. For example, the economic parameters may be represented by fuzzy numbers or intervals. And the stream conditions may exhibit stochastic uncertainties. These uncertainties should be tackled by appropriate inexact optimization methods. Maqsood et al.[1] proposed a method for water resources system planning by combing interval and fuzzy uncertainties into a two-stage stochastic programming. A new multi-purpose method has been presented by Khatiri et al.[2] in order to reach a compromise among multiple decision makers through identifying hydro-environmental management and policies. In this study, a two-stage inexact programming with value-at-risk (TIPV) model would be proposed for revealing dual uncertainties and reflecting the risks of recourse. A case study of TIPV on water resources management would be an attempt to combine the value-at-risk (VaR), the fuzzy boundary intervals and the random variables into a general two-stage programming framework.

2 Two-stage Inexact Programming with VaR

A two-stage inexact programming with VaR (TIPV) model is developed by introducing VaR into a two-stage inexact programming. TIPV can reveal uncertainties expressed intervals and probabilities in objective function and constraints. Moreover, possible risk loss of net system benefit could be measured.

*Corresponding author: 101582@bpi.edu.cn

$$\text{Max} \quad \tilde{f}^{\pm} = \sum_{i=1}^{m} \tilde{B}_i^{\pm} T_i^{\pm} - \sum_{i=1}^{m} \tilde{C}_i^{\pm} \left(\sum_{j=1}^{n} p_j S_{ij}^{\pm} \right), \tag{1}$$

s.t.

$$\sum_{i=1}^{m} \left(T_i^{\pm} - S_{ij}^{\pm} \right) \le q_j^{\pm}, \quad \forall j, \tag{2}$$

$$\xi_{FVaR}^{\pm} \left(\sum_{i=1}^{m} \tilde{C}_i^{\pm} \left(\sum_{j=1}^{n} p_j S_{ij}^{\pm} \right), \beta \right) \le \tau, \tag{3}$$

$$S_{ij}^{\pm} \le T_i^{\pm} \le T_{i\max}^{\pm}, \quad \forall i, j, \tag{4}$$

$$S_{ij}^{\pm} \ge 0, \quad \forall i, j. \tag{5}$$

Where $\tilde{f}^{\pm}$ is system benefit represented by a fuzzy boundary interval; $\tilde{B}_i^{\pm}$ and $\tilde{C}_i^{\pm}$ are benefit and penalty, respectively; $T_i^{\pm}$ is water allocation in first-stage; $S_{ij}^{\pm}$ is water shortage in second-stage; $q_j^{\pm}$ is available water resources under probability p_j; τ is maximum acceptable loss under probability β; $T_{i\max}^{\pm}$ is maximum allowable water allocation; $i = 1, 2, ..., m$ is index for water users; $j = 1, 2, ..., n$ is index for probabilities.

For decision makers, water resources could be promised to multiple demanders based on the requirement and the expected amounts of water resources in the first-stage. However, the inexact amounts of steam flow may bring out the water shortage problems. Therefore, economic penalties usually happen in the second-stage. TIPV is an attempt to combine the value-at-risk, fuzzy boundary intervals and random variables into a general two-stage programming framework. Economic penalties in the second-stage could be qualified and controlled by the VaR constraint in TIPV. That is to say, the optimal water resources allocations could be generated through maximum system benefits under controllable economic penalties.

The solution algorithm of TIPV could be developed based on two-step and VaR transformation methods. The two-step method is a useful method for solving two-stage programming[3]. Therefore, the difficulty of solution algorithm is the transformation of VaR constraint. Let $\tilde{\xi}^{\pm}$ be intervals with bounds of trapezoidal fuzzy random variable. The inexact VaR function $\xi_{FVaR}^{\pm}\left(\tilde{\xi}^{\pm}, \beta\right)$ are formulated as[3,4]:

$$\xi_{FVaR}^{\pm}\left(\tilde{\xi}^{\pm}, \beta\right) = \begin{cases} a_{\xi}^{\pm} + 2\left(b_{\xi}^{\pm} - a_{\xi}^{\pm}\right)\beta, & \text{if } \beta \le 0.5, \\ 2c_{\xi}^{\pm} - d_{\xi}^{\pm} + 2\left(d_{\xi}^{\pm} - c_{\xi}^{\pm}\right)\beta, & \text{if } \beta > 0.5. \end{cases} \tag{6}$$

3 Application in Water Resources Management

A case study on water resources management has been illustrated for application of the proposed TIPV method. There is an unregulated reservoir in this study. It supplies water resources to three water users (i.e. a municipality, an agricultural sector and an industry concern). Table 1 gives the economic data, including benefit and penalty. Table 2 gives the stream flow data. The related data about water allocation targets is shown in Table 3.

 https://doi.org/10.1051/e3sconf/202234101020

Table 1. Economic data ($/m^3).

	Net benefit $\tilde{B}_i^{\pm}$	Penalty $\tilde{C}_i^{\pm}$
Municipal $(i=1)$	[(83, 86,89, 92), (101, 104, 107, 110)]	[(211, 215, 219, 223), (270, 273, 276, 279)]
Industrial $(i=2)$	[(40, 45, 50, 55), (65, 70, 75, 80)]	[(55, 58, 61, 64), (84, 87, 90, 93)]
Agricultural $(i=3)$	[(21, 22, 23, 24), (28, 29, 30, 31)]	[(43, 46, 49, 52), (65, 68, 71, 74)]

Table 2. Stream flows.

Flow level	Probability (%)	Stream flow $q_j^{\pm}$ (in10^6 m^3)
Low (L) $(j=1)$	0.3	[3.4, 4.2]
Medium (M) $(j=2)$	0.4	[7.7, 12.6]
High (H) $(j=3)$	0.3	[15.0, 18.6]

Table 3. Water allocation targets (10^6 m^3).

	Municipal $(i=1)$	Industrial $(i=2)$	Agricultural $(i=3)$
Water allocation target $X_i^{\pm}$	[1.5, 2.3]	[2.7, 3.6]	[3.0, 5.2]
Maximum allowable allocation $X_{i\max}^{\pm}$	8.0	8.0	8.0

Table 4 gives the solutions obtained from TIPV. There exist water shortages for industrial and agricultural water users under low level of stream flow. Only agricultural water user's water demand could not be satisfied under medium level of stream flow. And under high level of stream flow, all water demand from three water users could be satisfied. Moreover, the amount of water shortage decreases as the risk level increases. It means that the VaR constraint for penalty has an effective on risk management.

Table 4. Solutions obtained from TIPV (10^6 m^3).

Risk level β	Shortage $Y_{ij}^{\pm}$								
	$Y_{11}^{\pm}$	$Y_{21}^{\pm}$	$Y_{31}^{\pm}$	$Y_{12}^{\pm}$	$Y_{22}^{\pm}$	$Y_{32}^{\pm}$	$Y_{13}^{\pm}$	$Y_{23}^{\pm}$	$Y_{33}^{\pm}$
0.5	0	[1.70, 2.50]	5.11	0	0	[0,3.31]	0	0	0
0.7	0	[1.70, 2.50]	4.54	0	0	[0,2.74]	0	0	0
0.9	0	[1.70, 2.50]	4.39	0	0	[0,2.59]	0	0	0
0.95	0	[1.70, 2.50]	4.35	0	0	[0,2.54]	0	0	0

Table 5 gives the results of system benefits under various risk levels. It indicates that the lower bound of system benefit increases as the risk level increases, and the upper bound of system benefit decreases as the risk level increases. That is to say, the increasing risk level would lead to an decreasing of interval width.

Table 5. Net system benefit ($ 10^6).

Risk level β	Distribution of net system benefit $\tilde{f}^{\pm}$
0.5	[(193.46, 227.98, 231.83, 251.02), (515.30, 539.18, 563.06, 586.94)]
0.7	[(207.47, 255.21, 260.26, 280.65), (506.71, 530.53, 554.36, 578.18)]
0.9	[(211.24, 262.54, 267.91, 288.62), (504.51, 528.32, 552.13, 575.94)]
0.95	[(212.32, 264.63, 270.10, 212.32), (503.98, 527.78, 551.59, 575.39)]

Figure 1 shows the optimal water allocation targets under β= 0.5, 0.7, 0.9 and 0.95. There always exist water shortage for water user of agriculture section under different stream flow levels under each risk levels. That is because the net benefit from agriculture section is the lowest among three water users. Moreover, comparing with other water users, the penalty for agriculture section is the lowest. Figure 2 shows the lower bound of optimal water allocation amounts. Figure 3 shows the corresponding upper bound. Results could supply multiple strategies for decision makers in water resources system planning.

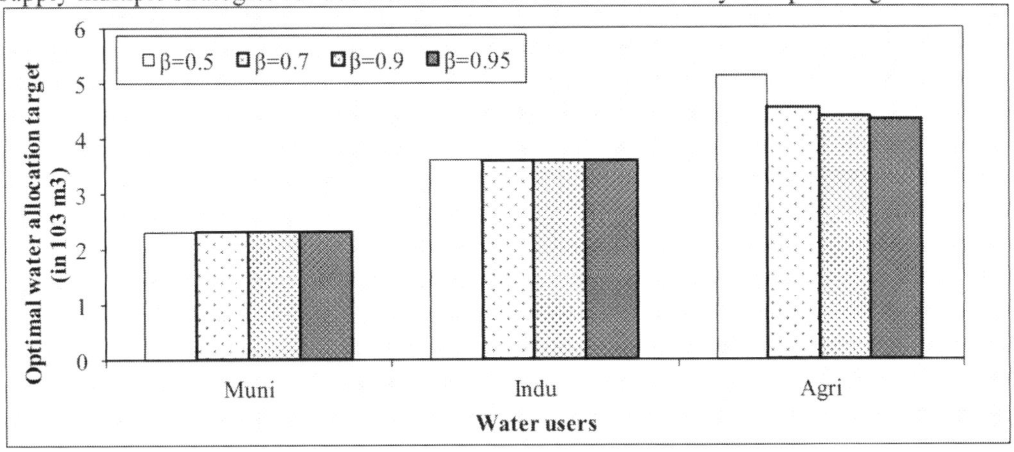

Fig. 1. Optimal water allocation targets.

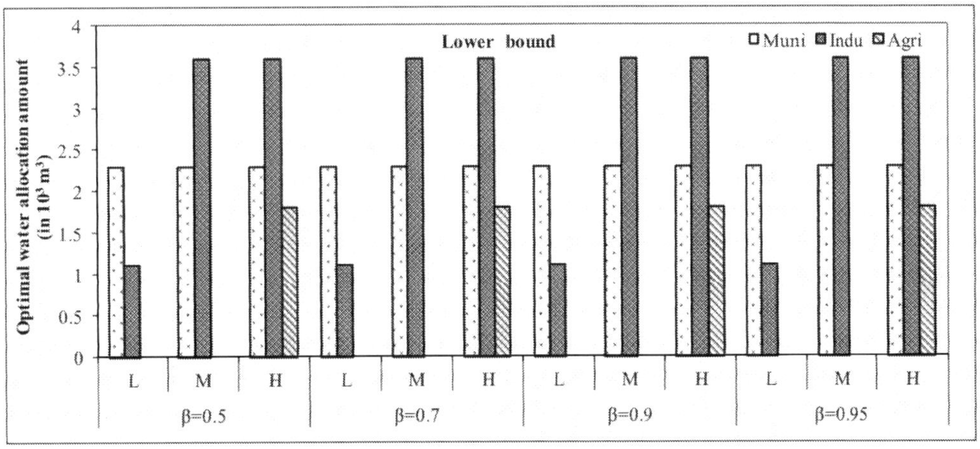

Fig. 2. Lower bound of the optimal water allocation amount.

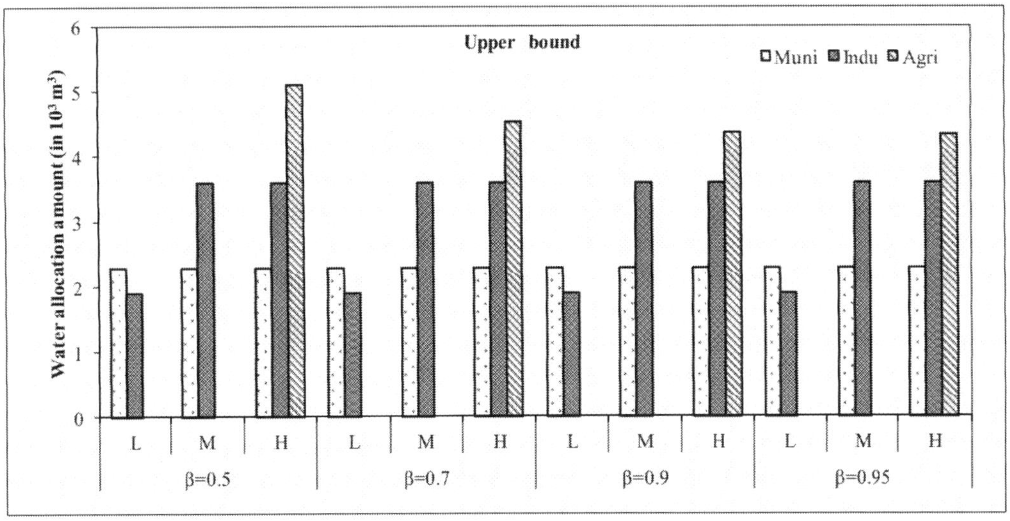

Fig. 3. Upper bound of the optimal water allocation amount.

4 Concluding Remarks

A case study on water resources management has been illustrated for application of a two-stage inexact programming with VaR (TIPV) method in this study. The advantages of TIPV are that (1) it could deal with inexact parameters presented as fuzzy boundary intervals and probability distributions; (2) it could reveal the potential risk of recourse at the second-stage. The results of case study show that more alternatives under various risk levels could be generated. The width of system benefit intervals would decrease as the risk level increases. Once it appears water shortage, the promised water allocation to agriculture section would be reduced because of the lowest net benefit and penalty of agriculture section among three water users. However, the amount of water shortage would decrease as the risk level increases. It means that the VaR constraint for penalty has an effective on risk management.

References

1. Maqsood I., Huang G. H., Yeomans J. S. (2005). An interval-parameter fuzzy two-stage stochastic program for water resources management under uncertainty. European Journal of Operational Research, 167(1):208-225. http://doi.org/10.1016/j.ejor.2003.08.068
2. Khatiri K. N., Niksokhan M. H., Sarang A., Kamali A. (2020). Coupled simulation-optimization model for the management of groundwater resources by considering uncertainty and conflict resolution. Water Resources Management, 34(11):3585-3608. http://doi.org/10.1007/s11269-020-02637-x
3. Kong X. M., Huang G. H., Fan Y. R., Li Y. P., Zeng X. T., Zhu Y. (2017). Risk analysis for water resources management under dual uncertainties through factorial analysis and fuzzy random value-at-risk. Stochastic Environmental Research & Risk Assessment, 31:2265–2280. http://doi.org/ 10.1007/s00477-017-1382-3
4. Shiraz R.K., Tavana M., Fukuyama H. (2020). A random-fuzzy portfolio selection DEA model using value-at-risk and conditional value-at-risk criteria. Soft Computing, 24:17167–17186. http://doi.org/10.1007/s00500-020-05010-7

Energy-saving and influencing factors of household electricity consumption in China

Ming Wen[1,2], *Yun* Zhang[1,2], *Hao* Xiao[3], and *Rui* Huang[3,*]

[1] Hunan Key Laboratory of Energy Internet Supply-demand and Operation, 410004 Changsha, China
[2] State Grid Hunan Electric Power Company Limited Economic & Technical Research Institute, 410004 Changsha, China
[3] Hunan University, School of economics and trade, 410079 Changsha, China

Abstract. Energy conservation and environmental protection are crucial to green development. This article distinguishes the relationship between the electricity literacy, energy saving awareness and information feedback. We found that i) electricity literacy and information feedback are negatively correlated with electricity consumption, ii) there is a positive correlation between electricity consumption and living area, household income, permanent resident population and education level, iii) electricity consumption presents an inverted U shape with education level.

1 Introduction

Energy conservation and environmental protection is one of China's basic state policies and has been the focus of the whole society. According to the report of the 19th CPC National Congress, "To promote green development, we must promote comprehensive conservation and recycling of resources, promote a revolution in energy consumption and reduce energy consumption." [1] In addition to technological progress, improving the consciousness of saving of the whole society is undoubtedly an important way to save resources. Therefore, changing the concept of residents' energy consumption and reducing energy waste has become an important breakthrough to solve the energy problems [2].

At present, many scholars have done a lot of research on electric power energy, mainly in the following aspects: 1) related research on energy conservation awareness. Hori et al. (2013) [3] and Wang et al. (2018) [4] found that community and individual awareness of energy conservation will affect daily energy conservation activities. Lee (2015) [5] and Abrahamse et al. (2009) [6] found that students who master more electricity knowledge are more able to make reasonable energy consumption behavior. 2) Studies on the influencing factors of energy consumption. In terms of age and family income, Dirk et al. (2012) [7] found that the aging of population and the increase of family income were not conducive to the promotion of energy conservation. In terms of energy price and housing conditions, energy price and living area have a significant impact on household energy consumption expenditure, and rural household energy consumption depends on the price of electricity [8].

*Corresponding author : ruihuanghr@126.com

To sum up, the existing literature has some limitations in studying the relationship between energy conservation awareness and energy consumption behavior. This is mainly reflected in the lack of empirical analysis of the differences in energy consumption behavior based on large-scale questionnaire survey data in China. Compared with previous studies, the contributions of this article include three aspects: Firstly, we innovatively divide the main factors affecting electricity consumption into three aspects, including electricity literacy, awareness and feedback. Secondly, to observe the influence of various factors on electricity consumption behavior under different income levels. Thirdly, we find that education level and electricity consumption show an inverted U shape, which increases first and then decreases with the improvement of education level.

2 Model setting and data source

Based on the literature survey above, the following measurement model was established to investigate the impact of energy conservation awareness indicators:

$$\ln energy = \beta_0 + \beta_1 awareness + \beta_2 feedback + \beta_3 literacy + \beta_4 X + \varepsilon \tag{1}$$

Literacy indicates the degree to which people regard energy literacy. In this paper, the energy efficiency labels of the main electrical appliances (refrigerators) (from grade1 to 5) were selected as the measurement index, and set the five dummy variables. Awareness indicates the level of people know about energy. In this paper, "whether the policy of stepped electricity price is known" is selected as its measurement index and set as a dummy variable (if the stepped electricity price is known, the value is 1). Feedback means information feedback. The measurement index is "frequency of electricity payment" (if the electricity payment is paid once a month, the value is 1). X represents other control variables. ε represents error term.

The control variables include city, age, gender, education level, actual living area of residents, annual income of residents, permanent household population, electricity price, etc. City represents the urban-rural dummy variable (defined as 1 for city and 0 for rural). Old represents the age dummy variable (>60 years old is 1). Female represents gender (female is 1). Educate stands for education level, including five categories (uneducated, elementary, middle, high and college degree or above). Area is the actual living area. Income means annual income. POP is the number of people living in a household. Price represents the average price of electricity, which is calculated by dividing the monthly cost of electricity by the amount used.

The data are from the 2015 Chinese General Social Survey (CGSS2015). It adopted multi-stage stratified sampling method, and 10,968 valid questionnaires were completed from 478 villages in 28 provinces, municipalities and autonomous regions.

Table 1. Descriptive statistics.

variable	Observed	Mean	S.D.	Min	Max
Electricity consumption	2857	4.577	0.844	0	8.006
literacy	1899	1.635	0.865	1	5
awareness	3489	0.383	0.486	0	1
feedback	3489	0.644	0.479	0	1
old	10809	50.333	16.849	18	95
female	10809	0.531	0.499	0	1
educate	10780	2.992	1.267	1	5
city	10809	0.588	0.492	0	1
income	9718	10.508	1.149	5.298	16.118
pop	10776	0.945	0.499	0	2.565

area	10754	4.544	0.633	2.303	6.397
price	3073	-0.187	0.791	-3.188	4.748

3 The empirical analysis

3.1 The Baseline Regression

Based on the model (1), this paper obtained the following basic regression results. As can be seen from Table 2, we can include four aspects: firstly, when the energy label of refrigerators represents energy literacy, it means they have higher energy literacy, have higher awareness of energy conservation, and can use energy more rationally. Secondly, when the feedback frequency is higher, the electricity consumption is less, that is, every 1% increase in the feedback frequency will reduce the electricity consumption by about 0.06%. Thirdly, urban residents use more electricity than rural ones. In terms of education, it may be that high level of education is often associated with high income, for every 1% more urban residents, electricity consumption increases by about 0.3%. Fourthly, education level and electricity consumption show an inverted U shape, which increases first and then decreases with the improvement of education level. Otherwise, residential area, household income, household permanent resident population and average monthly electricity price have positive effects on household electricity consumption.

Table 2. Baseline regression result.

	(1)	(2)	(3)	(4)
grade1	-0.355***	-0.352***	-0.257**	-0.250**
	(-2.89)	(-2.75)	(-2.00)	(-2.23)
grade2	-0.364***	-0.365***	-0.286**	-0.283**
	(-2.91)	(-2.81)	(-2.17)	(-2.48)
grade3	-0.423***	-0.422***	-0.366***	-0.370***
	(-3.18)	(-3.05)	(-2.66)	(-3.04)
grade4	-0.421*	-0.481**	-0.361	-0.229
	(-1.88)	(-2.10)	(-1.64)	(-1.38)
Awareness		0.294***	0.180***	0.155***
		(7.72)	(4.71)	(4.23)
Feedback		-0.078*	-0.066*	-0.027
		(-1.88)	(-1.68)	(-0.71)
female			-0.020	-0.014
			(-0.54)	(-0.39)
old			0.051	0.054
			(1.16)	(1.28)
city			0.364***	0.286***
			(8.02)	(6.18)
edu2			0.158**	0.094
			(2.28)	(1.40)
edu3			0.238***	0.116*
			(3.48)	(1.74)
edu4			0.304***	0.179**
			(4.00)	(2.40)
edu5			0.281***	0.108
			(3.52)	(1.36)
area			0.158***	0.127***
			(4.60)	(3.65)
pop			0.374***	0.270***
			(9.25)	(6.62)

	(1)	(2)	(3)	(4)
income				0.147***
				(7.21)
price				-0.303***
				(-11.53)
Observations	1584	1584	1575	1455
R-squared	0.004	0.042	0.162	0.285

legend: * p<0.1, ** p<0.05, *** p<0.01

3.2 Further test

In the analysis of the regression model and in connection with the questionnaire, we found that income had a certain inhibiting and promoting effect on the ability to influence variables such as energy literacy and energy conservation knowledge. To discuss and verify this effect, this paper performed a piecemeal test of income. Four regression models with 25%, 50% and 75% as demarcation points were constructed. Relevant data description is shown in Table 3.

Table 3 shows that energy literacy and feedback have different influence on energy consumption in the case of different income levels. When the annual income is at the low income level, energy literacy has a greater impact on energy consumption among low-income groups. Under different income levels, the influence coefficient of energy saving knowledge on energy saving effect is different. It can be considered that this process is influenced by income and has a certain endogeneity. At the same time, the influence of awareness on electricity consumption is significantly positive and increases with the increase of income. We guess that low-income groups tend to pay more attention to the change of electricity price, which will significantly change their electricity consumption habits. However, high-income people pay less attention to electricity price, and the change of electricity price has little effect on their electricity consumption habits.

Table 3. Grouped regression analysis based on income.

Income%	0—25% (low income)	25%—50%	50%—75%	75%—100% (high income)
grade_1	-0.7987**	-0.0546	-0.1340	-0.4113***
grade_2	-1.0590***	-0.0546	-0.0988	-0.4233***
grade_3	-0.6335	-0.3341**	-0.0838	-0.4888***
grade_4	-1.3031***	-1.2076***	-0.1616	-0.0950
awareness	0.0909	0.1505**	0.1686**	0.1789***
feedback	-0.0841	0.0944	-0.0915	-0.1273*
female	-0.0663	0.0011	-0.0281	-0.0103
old	0.1244	0.1291	0.0010	0.0000
city	0.4870***	0.1706**	0.2662***	0.3774***
edu2	0.2117*	-0.0137	0.1774	0.2963
edu3	0.2283*	0.1352	0.1942	0.1415
edu4	0.4073**	0.1268	0.2194	0.1766
edu5	0.5182***	0.0310	0.1525	0.0402
area	0.0538	0.0521	0.1725**	0.0843
pop	0.3672***	0.3221***	0.2152***	0.3297***
price	-0.3440***	-0.2944***	-0.3436***	-0.2846***
cons	4.2667***	3.7174***	3.3615***	4.2217***

legend: * p<.1; ** p<.05; *** p<.01

4 Conclusion and suggestion

Based on the CGSS2015 data, this paper conducts the influence factors of electricity literacy, energy saving awareness and information feedback. We found that the higher the electricity literacy, the better the electricity efficiency. With the improvement of people's awareness of electricity saving, the effect of electricity saving is more obvious. We also found that the more frequently pay the electricity bill, the less electricity you use. In addition, education level and electricity consumption show an inverted U shape. So, we will guide the society to create conditions for energy conservation and cultivate people's energy conservation habits. Then detailed policies to different groups should be made to promote the diversification of energy conservation policies. Finally, we should refine the content of information feedback and innovate the method of information feedback.

References

1. The State Council, Strategic action plan for energy development, 6, (2014)
2. The State Council, The 13th Five-year Plan for energy developmen, 12, (2016)
3. Hori, S., Kondo, K., Nogata, D., & Ben, H., Energy Policy, 52(2013)
4. Wang, B., Wang, X., Guo, D., Zhang, B.,& Wang, Z., Energy Policy, 116,(2018)
5. Lee, L. S., Lee, Y. F., Altschuld, J. W., & Pan, Y. J. , Energy Policy, 76, 76(2015)
6. Abrahamse, W., & Steg, L., Journal of Economic Psychology, 30,5(2009)
7. Brounen,D., Kok,N., & Quigley,J. M., European Economic Review, 56, 5(2012)
8. Costa, D. L., & Kahn, M. E., American Economic Review, 101, 3(2011)

Calculate of hunan industrial early warning index

Lang Ren[1], *Dan* Li[2], *Junlin* Zhou[3,*], and *Yongli* Liu[3]

[1] State Grid Hunan Zhangjiajie power supply company, Zhangjiajie, China
[2] State Grid Hunan Yiyang power supply company, Yiyang, China
[3] School of economics and trade,Hunan University, Changsha, China

Abstract. Electric Power data has the characteristics of strong real-time, fine granularity and high accuracy. It can more accurately reflect the current industrial structure and is suitable for the monitoring of high-quality economic development. In order to better play the role of power data monitoring and early warning,based on the prosperity signal lamp method, combining the principle of mathematical probability method, the early warning index system of electric power economic monitoring is constructed. The results show that the index can monitor the power situation of Hunan Province, and can be used in the construction of green energy and promote the energy conservation and high-quality economic development of Hunan Province.
Keywords: Power economy early warning index; Early warning monitoring; Hunan Province.

1 Introduction

Electric power demand is closely related to economic and social development. Electric power demand is regarded as the "barometer" and "thermometer" of macro-economy. Electric power demand involves all aspects of economic operation and people's life. In view of the accuracy of data collection, electric power data can more truly reflect the economic operation compared with traditional macroeconomic data (Tan et al. 2016[1]).

The index construction method has always been one of the important means for domestic and foreign scholars to judge the macro and industrial development. The main methods of macroeconomic early warning include prosperity index method and prosperity signal lamp method. From the perspective of macroeconomic early warning, Yu and Zhang (2020)[2] used the idea of signal light to build a foreign exchange reserve risk system in order to effectively monitor foreign exchange risk; Liu (2021)[3] constructed the early warning framework of China's Urban Smart supply chain based on the improved signal lamp model. Give early warning information according to the status of the signal light, and effectively provide suggestions for the development of smart supply chain for government departments; Yan et al. (2021)[4] monitored and warned the macroeconomic operation based on support vector machine.

* Corresponding author: 2534331171@qq.com

From the perspective of the industry early warning system, Shi and Wang (2017)[5]introduced the prosperity analysis into the electric power market, and compiled the leading index of electric power prosperity, predicted the turning point of future electric power demand development and speculated the future trend of electric power demand by observing the index; Chen and Wang (2019)[6] analyzed the fluctuation of service industry by constructing a monthly consistent prosperity index of China's service industry through six service industry related indicators. The results show that the growth of China's service industry generally presents the characteristics of "slow growth and rapid decline" in the 21st century, which improves the early warning research of the service industry; Zhou et al. (2021) [7]built an MF-MS-SW model that can analyze seasonal and monthly frequencies at the same time, selected the leading, consistent and lagging mixing sample data composed of 21 indicators, and built China mixing electric power prosperity index and early warning signal system. Sun (2021)[8] proposed and constructed a regional electric power prosperity early warning system based on multi-dimensional big data such as economy, energy, electric power and environment on the basis of analyzing common electric power prosperity indexes and early warning methods; However, the research on early warning index mostly starts from the macro-economy, so it often only includes various macroeconomic indicators in the selection of indicators, or studies the early warning system of different industries. Few studies start from the correlation between electric power and economy, look at economic development with electric power, and force electric power with economy. This paper starts from the two aspects of electric power and economy, The early warning index of Hunan electric power economy is constructed to monitor the economic development of Hunan Province.

2 Construction of early warning index of electric power economy in Hunan Province

The design of prosperity signal system is based on the selection of advance and consistent system. The specific process includes the selection of signal indicator, the determination of indicator threshold and weight, and the calculation of comprehensive alarm index.

The early warning index in this paper is used to measure the economic prosperity of Hunan. It needs to use all the indicators in Hunan Province, covering the main aspects of Hunan economy. According to the data indicators selected by China to establish the early warning index and the economic development of Hunan Province, the index system of Hunan macroeconomic early warning index is finally determined, as shown in table 1, with 15 data in total.

Table 1. Early warning index system.

Order	Name of Index
	Economic Indicators
X_1	Growth of industrial added value
X_2	Money supply (M1) increases
X_3	Loan balance growth in local and foreign currencies
X_4	Fixed asset investment growth rate
X_5	Growth rate of total retail sales of consumer goods
X_6	Import growth
X_7	Export growth
X_8	Growth in fiscal revenue
X_9	Growth in consumer price
X_{10}	Real estate investment growth
	Power Data

Y_1	Industrial electricity consumption was used year on year
Y_2	Ferrous metal smelting and rolling processing industry
Y_3	Manufacture of non-metallic mineral products
Y_4	Transportation, electrical, and electronic equipment manufacturing
Y_5	Production and supply of electricity, gas and water

We use mathematical statistics to determine the boundary of early warning indicators and divides the boundary of early warning indexes according to the specific steps of:

(1) The score is determined by determining a single warning indicator critical value. Note: Sample mean and sample standard deviation σ were calculated from the sample data of each boom warning indicator. The critical values of the warning index are shown in Table 2. Then, according to the interval of indicator value in each period, the signal light color of each indicator in each period is judged. Each signal is given different points, 5 for "red", 4 for "yellow", 3 for "green", 2 for "light blue" and 1 for "dark blue".

Table 2. Cut-off value of Hunan electric power economic early warning index.

Metrics	**Critical Value**			
	red←yellow ($u+2\sigma$)	**yellow←green** ($u+\sigma$)	**greenlight←blue** ($u-\sigma$)	**lightblue←blue** ($u-2\sigma$)
X_1	9.998	8.68	6.06	4.742
X_2	4.302	3.20	1.00	-0.102
X_3	18.84	14.49	5.79	1.44
X_4	32.77	20.10	-5.24	-17.91
X_5	106.06	68.61	-6.29	-43.74
X_6	77.57	51.03	-2.05	-28.59
X_7	40.35	29.26	7.08	-4.01
X_8	24.19	19.70	10.72	6.23
X_9	26.33	18.55	2.99	-4.79
X_{10}	18.46	16.95	13.93	12.42
Y_1	24.72	14.61	-5.61	-15.72
Y_2	11.31	5.90	-4.92	-10.33
Y_3	35.21	20.24	-9.70	-24.67
Y_4	46.7	26.36	-14.32	-34.66
Y_5	46.43	26.24	-14.14	-34.33

(2) Determine the valve value of the composite index score. After determining the score of the composite index, the boundary value corresponding to the composite index is determined, and then the signal light color corresponding to the composite index is determined. The formula is below. The specific results are shown in Table 3.

$$\left.\begin{array}{l} I_{\text{yellow}\rightarrow\text{red}} = 15 \times 5 \times 85\% = 63.75 \\ I_{\text{green}\rightarrow\text{yellow}} = 15 \times 5 \times 74\% = 55.5 \\ I_{\text{light blue}\rightarrow\text{green}} = 15 \times 5 \times 50\% = 37.5 \\ I_{\text{blue}\rightarrow\text{light blue}} = 15 \times 5 \times 36\% = 27 \end{array}\right\} \quad (1)$$

Table 3. Threshold setting of Hunan electric power economic early warning index.

Warning status	**superheat**	**Hot**	**stabilize**	**Cold**	**supercooling**

signal lamp	Red	Yellow	Green	Light blue	Blue
threshold	>63.75	[55.5,63.75]	[37.5,52.25]	[27,37.5]	<27

(3) Determine the color of the comprehensive index signal light. See (2) for constructing comprehensive index targets.

$$Y_{ti} = Y_{t1} + Y_{t2} + Y_{t3} + Y_{t4} + Y_{t5} + Y_{t6} + Y_{t7} + Y_{t8} + Y_{t9} + Y_{t10} + Y_{t11} + Y_{t12} + Y_{t13} + Y_{t14} + Y_{t15} \quad (2)$$

Among them, Y is the comprehensive index score, t=1,2, … 60, representing different periods, i=1,2, … 15, representing 15 early warning indicators, Y_{t1}, Y_{t2}, Y_{t3}, Y_{t4}, Y_{t5}, Y_{t6},… Y_{t15}, representing the signal light score of 15 early warning indicators in period t, adding them to indicate the Hunan early warning index score.

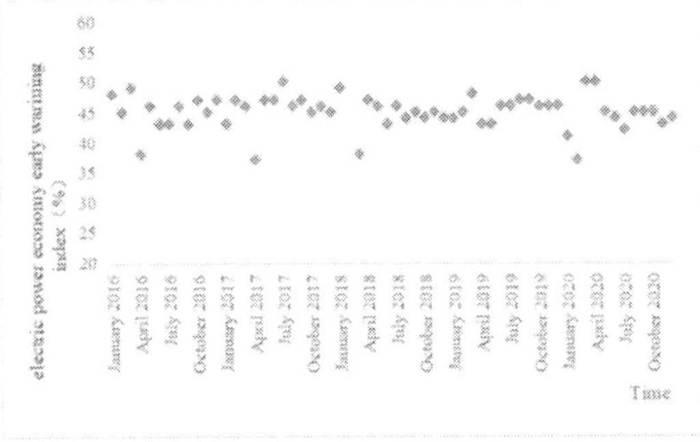

Fig.1. Trend chart of electric power economy early warning index in Hunan Province.

Hunan province electric power economy early warning index is mainly in green state, indicating Hunan province economic development is in normal state, the current economic and electric power development is very stable. As can be seen from Figure 1, the warning index of Hunan province was stable in 2016. In April 2016, it changed from the green light area to the light blue light area and then fell back to the green light area, indicating that the economic development of Hunan Province changed from normal to cold and then normal operation. According to the analysis of various indicators in April 2016, industrial electricity consumption and non-metallic mineral products industry electricity consumption are in the "light blue light" area, while the electric power consumption of transportation, electrical and electronic equipment manufacturing industry and electricity, gas and water production and supply industry are in the "blue light" area. On the one hand, it shows that Hunan electric power plays an important guiding role in economic development, and the economic development of Hunan Province needs the drive of electric power; On the other hand, except for the real estate industry and import and export growth in Hunan Province, other economic indicators are in the "green light zone " or "yellow light zone ", and the economic development of Hunan Province is relatively stable. Subsequently, the signal light of Hunan Province turned green after May 2016, and the early warning index rebounded and entered the "yellow light zone" in February 2018. The economy in 2018 became hot because of the influence of the growth of Hunan province's export and local fiscal revenue. These two indicators entered the "red light zone" in February 2018; influenced by the Spring Festival effect, the growth rate of social consumer goods in Hunan province is in a hot state; the growth rate of power consumption in ferrous metal smelting and calendering industry in Hunan province increased in February 2018, indicating that the economy in 2018 was mainly influenced by fiscal policy, export and holiday effect. In February 2020, the electric power economic index of Hunan Province fell into the "light blue light zone", and the electric power economy was cooling down, mainly due to the cooling of the electric power market and the impact of the decline in export and fiscal revenue growth rate. In general, the economic growth of Hunan Province during 2016-2020 has achieved stable growth in the safe and stable zone. Although the economic growth is hot, it has not reached the safety warning line, and the overall safety is achieved.

3 Conclusions and recommendations

Judging the economic development of Hunan Province according to the prosperity signal, it can be seen that Hunan Province is mainly under the condition of green light(In April 2016, the electric power economy early warning index entered the light blue light zone, in February 2018, the electric power economy early warning index entered the "yellow light zone", and in February 2020, the electric power economy early warning index of Hunan Province fell into the "light blue light zone"), The main reason for the change of Hunan electric power economic early warning index is the change of Hunan electric power economy and the impact of finance, import and export, but on the whole, Hunan's economy is stable. Based on the above research, the following suggestions are put forward for the development of Hunan Province:

(1) Ensure the stability of the electric power market and meet the economic development and consumer demand of residents.

(2) With high-quality economic development to force electricity, change the household electricity consumption pattern to improve the quality of life.

References

1. Y. M. Tan, W. Q. Sun, *Power Grid and Clean Energy*, **32 (07)** , 45-50 + 64(2016)
2. M. Yu, K. Zhang, *Management Review*, **32 (7)**,54-65(2020)
3. W. H. Liu, J. H. Hou, C. L. Yuan, W. F. *Wu, Industrial Technology and Economy*, **40 (2)**,56-64(2021)
4. Y. Yan, Y. Cheng, X. H. Sun, Statistics and Decision Making, **37 (14)**,137-140(2021)
5. L. Shi, Y. Wang, *Enterprise Management*, **(S2)**,356-358(2017)
6. Y. Z. Wang, L. Chen, Y. G. *Meng, Statistics and Decision Making*, **35 (21)**,30-34(2019)
7. D. C. Zhou, J. J. Chen, B. Liu, *China Power*,1-13
8. R. H. Sun， M. Yang, H. K. Bai, *Power Big Data*, **21 (09)**,52-60(2018)

The consistency measure of hunan's electric power industry prosperity index

Chufan Xu[1,2], and *Ming* Wen[12,*]

[1] State Grid Hunan Electric Power Company Ltd. Economic and Technical Reasearch Institute, Changsha , China
[2] Hunan Key Laboratory of Energy Internet Supply-demand and Operation , Changsha , China

Abstract. In view of the problem that the traditional electric power prosperity index lacks the support of the mixed frequency and multi-dimension data, a new consistency measure method is constructed, which uses time difference correlation analysis and principal component and consider the linkage of electric power and economy at the same time. Take the Hunan province for example, we set the pool of kinds of sectional electricity power consumption and other relatives data, use the method of principal component analysis to capture main character of electricity's change. The results show that the proposed consistency measurement model can describe the fluctuation characteristics of Hunan's Power Prosperity Index well. The model can be used for monitor the changes in the power industry status and promoting the policy-making about the adjustment of the energy-saving and environment-friendly structure.

1 Introduction

Electric power, as the "wind vane" of economic development, is closely related to economic development. Electrical data can more truly reflect the overall picture of economic development. According to the index used to evaluate China's GDP growth launched by the famous British political and economic magazine The Economist in 2010, electric power consumption can more truly reflect the overall picture of economic development. Compared with economic macro data, electricity consumption as the actual consumption is easier to verify, the statistical error is smaller, and it is more difficult to deliberately distort and falsify, so it can more accurately reflect the current economic situation.

Index construction method has always been one of the important means for domestic and foreign scholars to evaluate macroeconomic development[1-4]. Based on the synthetic index method, Fang [5]uses financial, order, expectation and other indicators to construct China's macroeconomic leading index, so as to better judge the macroeconomic situation. Since the "Li keqiang index" proposed, electricity consumption has gradually been widely concerned. As an important part of Li keqiang index, industrial electricity consumption can better reflect the overall picture of industrial economic development compared with pure economic indicators, so it has become one of the important indicators for scholars to

*Corresponding author: xflixian@163.com

evaluate economic development. Shi et al. [6] and Sun et al.[7] introduced the prosperity index model in the field of economics into the field of electric power market analysis and prediction, took the electric power consumption of the whole society as the benchmark index. Liu et al. [8] constructed an economic prosperity index based on electric power big data, which objectively reflects the economic operation status, assists in forecasting the trend of economic development, and achieves the goal of observing the economy from the perspective of electric power. Zhou et al. [9] constructed an MF-MS-SW model capable of analyzing seasonal and monthly frequencies at the same time, and selected the antecedent, consistent and lagged mixing sample data composed of 21 indicators to construct China's mixed-frequency electric power prosperity index and warning signal system, so as to continuously evaluate and monitor China's electric power prosperity.

Most of the current studies build indexes through electric power data to predict electric power consumption, but ignore the role of electricity in predicting economy. To sum up, taking Hunan Province as an example, this paper introduces the prosperity analysis method into the electric power consumption market, constructs the electric power economic prosperity index, effectively monitors and warns the electric power in Hunan Province, and predicts the electric power economic situation in Hunan Province.

2 Index selection and data processing

The general influencing factors of power economic index selection and the operation characteristics of hunan province's macro-economy are considered comprehensively.In this paper, indicators are screened from both economic and electric aspects to construct the power economic indicator pool of Hunan Province, as shown in Table 1.

Table1. Alternative index pool of electric power economic index

Index classification	name of index
Consumption index	Hunan: Total retail sales of social consumer goods
Investment indicators	Hunan: Real estate development investment completed amount Hunan: fixed asset investment
Foreign trade indicators	Hunan: Export amount Hunan: Import amount
Financial indicators	Hunan: Local public finance revenue Hunan: Local public finance expenditure
Price indicators	Hunan:CPI Hunan:RPI Hunan:PPI
Monetary Capital Index	Hunan: Balance of local and foreign currency loans Hunan: Balance of local and foreign currency deposits
Industrial indicators	Hunan: Iron ore production Hunan: Raw iron production Hunan: crude steel output Hunan: Steel production Hunan: Cement output
Power data of key industries	Industrial electricity consumption Electricity consumption for smelting of black metal and rolling processing industry Electricity consumption of non-metallic mineral products industry

	Electricity consumption of chemical raw materials and chemical products manufacturing industry
	Electricity consumption of non-ferrous metal smelting and rolling processing industry
	Transportation, electrical, electrical equipment manufacturing electricity consumption
	Electricity consumption in the production and supply of electric power and heat industry
National indicators	Macroeconomic climate index: the consensus index
	Macroeconomic climate index: the leading index
	Macroeconomic climate index: lag index
	Export amount
	fiscal expenditure
	M2: on year-on-year basis

To determine the leading, consistent or lag time difference relationship of different indicators on the macro economy, it is first necessary to choose a benchmark reference index, as the standard to judge the nature of other indicators, as well as to test the results. When developing the monthly power economic index, we adopted the year-on-year growth rate of industrial added value as the benchmark index.

3 Empirical results

Time Difference Correlation analysis is a common method to verify the relationship among precedence, consistency and lag of time series by correlation coefficient. $Y = (y_1, y_2, \cdots, y_n)$ as a benchmark, $X = (x_1, x_2, \cdots, x_n)$ as the selected indicator, r is the Correlation Coefficient, so

$$r_l = \frac{\sum_{t=t'}^{T_l}(x_{t+l}-\bar{x})(y_t-\bar{y})}{\sqrt{\sum_{t=t'}^{T_t}(x_{t+l}-\bar{x})\sum_{t=t'}^{T_l}(y_t-\bar{y})^2}}, l=0,\pm1,\pm2,\ldots, t' = \begin{Bmatrix} 1, & l \geq 0 \\ 1-l, & l < 0 \end{Bmatrix} \quad (1)$$

In the formula, L is called the time difference or the delay number, representing the advance period or the lag period. When l takes a negative number, L is ahead; when l takes a positive number, L is behind; l is the maximum dela. T_l is the number of data after the data collection. In this paper, the data of the index pool is converted into the relative value data of the same month, and the Correlation Coefficient between the benchmark index and the index pool is calculated. When the number of lag period is greater than 3, x lag y index is denoted; When the lag is less than -3, X is ahead of Y; When the lag period number is [-3,3], x and Y are synchronized. In this part, the Correlation Coefficient of time difference in 20 periods of time lag is calculated, and the Correlation Coefficient of time difference of each index in the index pool is obtained.

The Correlation Coefficient of time difference with industrial value added in the same month was less than 0.3, including the amount of real estate development investment in Hunan Province, the electricity consumption of chemical raw materials and chemical manufacturing in Hunan Province, and the national fiscal expenditure.There are 24 indexes with time difference Correlation Coefficient above 0.3, among which 9 indexes with Correlation Coefficient above 0.3 and delay period number between (- 3,3) are consistent indexes,The number of delay periods below-3 was 7, which was identified as the leading indicator.There were 11 delay periods with more than 3, which were the index of delay.

This paper mainly constructs the power economic index based on principal component analysis. Firstly, the data are standardized, and then based on the standardized power and economic indicators, the number of components is screened by principal component analysis, and the power economic index of Hunan Province is calculated. The following figure shows the consistency index of Hunan electric power economy.The leading index and the lagging index is the same.

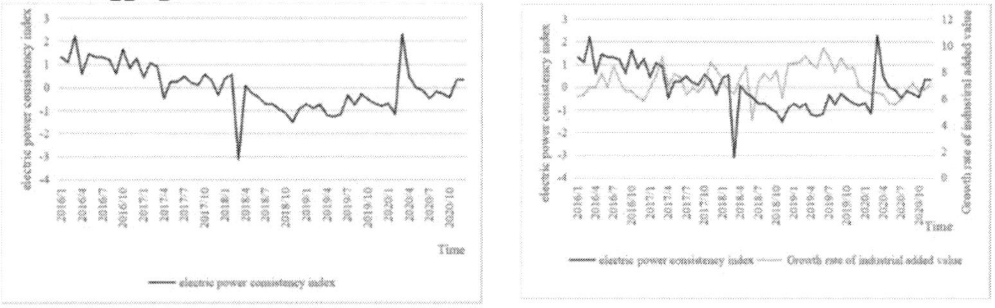

Fig. 1. Hunan Electric Power Economic consistency index.

Compare the constructed Hunan Electric Power Economic consistency index with the growth rate of industrial added value in Hunan Province, as shown in Fig.1. The growth rate of industrial added value in Hunan Province is roughly the same as that of Hunan Electric Power Economic consistency index. In addition to April 2018, Hunan Province was affected by the effect of the Spring Festival and the unsealing of the epidemic in April 2020. The consistency index can effectively monitor the economic development of Hunan Province.

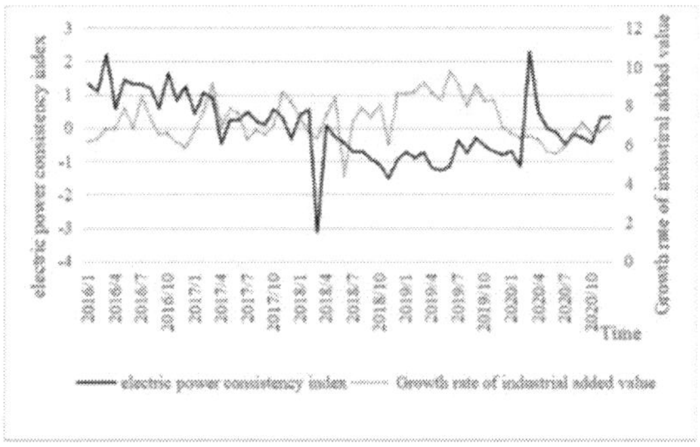

Fig. 2. Power economy consistency index and industrial added value trend of Hunan Province.

Compare the constructed Hunan electric power economy leading index with the growth rate of industrial added value in Hunan Province, as shown in Fig3. It can be found that from January 2016 to October 2016, the growth rate of industrial added value in Hunan Province is consistent with the leading index of electric power economy in Hunan Province; In November 2016, the leading index of Hunan electric power economy showed a peak, and the growth rate of leading industrial added value was 3 months; Meanwhile, the leading index of Hunan electric power economy in August 2017 is consistent with the growth trend of industrial added value in November 2017. Overall, the industrial added value of Hunan Province and the leading indicators of Hunan electric power economy show a fluctuating trend, but the leading indicators of power economy lead the industrial added value of Hunan Province for about three periods (three months). By constructing the leading indicators of Hunan electric power economy, we can effectively predict the economic growth trend of Hunan Province.

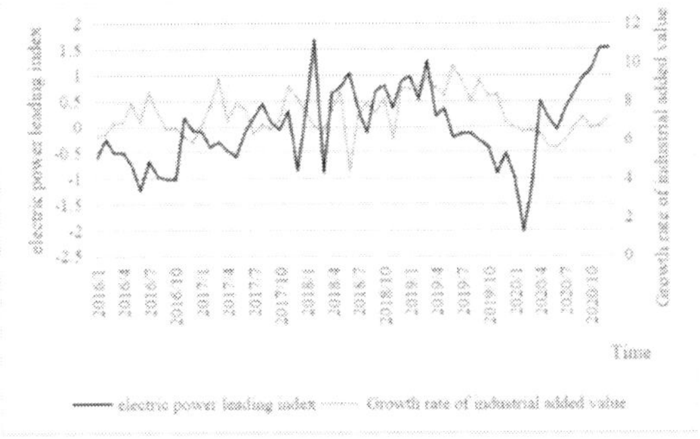

Fig. 3. Power economy leading index and industrial added value trend of Hunan Province.

4 Conclusions

According to the power economic index of Hunan Province, referring to the structure constructed by relevant scholars and Keqiang index, the driving force of economic development of Hunan Province and the power consumption of industrial industries with high power share are included in the index pool. Taking the year-on-year industrial added value of Hunan Province as the benchmark index, and according to the time difference correlation analysis, the leading index, consistent index and lagging index are selected in the index pool to construct the leading index, consistent index and lagging index of Hunan electric power economy. The constructed leading index of electric power economy leads the industrial added value for 3 months, and the constructed consistent index of electric power economy is synchronized with the industrial added value, The constructed index is tested. Therefore, the construction of Hunan electric power economic index can effectively evaluate the current economic development situation of Hunan Province.

Acknowledgements

State Grid Hunan Electric Power Co. , Ltd. Project Support (B616A221000J)
Science-Technology Innovation Pltaform and Talents Program of Hunan Provinve,China under Grant 2019TP1053

References

1. X.W. Li, M.X. Li, *System engineering*,**33**,12(2015)
2. Y. Hong, C.W. Ma, *Statistics and decision making*,**36**,13(2020).
3. J.S. Wang, Y.H. Ren, *Research on quantitative economy and technical economy*,**38**,2(2021).
4. W.H. Liu, W.F. Wu, C. L. Yuan, J. H. Hou, *Industrial technology and economy*,**40**,1(2020).
5. H.G. Fang, *World research*, **8**(2021).
6. L. Shi, Y. Wang , *Enterprise management*, **S2**(2007).
7. R. H. Sun , M. Yang, H. K. Bai, *Power big data*,**21**,9(2008).
8. Y.J. Liu, K.H. Song, X . Wang, *Telecommunications science*,**36**,6(2020).
9. D.C. Zhou , J.J Chen, B. Liu, *China power*.

Analysis and research on energy utilization efficiency of port portal crane

Leilei Liu[1*], *Kun* Guan[2], and *Peisen* Wu[1]

[1] Tianjin Research Institute for Water Transport Engineering, Key Laboratory of Environmental Protection Technology on Water Transport, Ministry of Transport, Tianjin, China
[2] Tianjin Port Zhongmei Huaneng Coal Terminal Co., Ltd

Abstract. Portal crane is one of the important equipment of bulk terminal, which is specially used for shore loading and unloading operation. It mainly consumes electric energy. In order to clarify the utilization of electric energy during the operation of portal crane and analyze the energy utilization efficiency, a portal crane of a bulk terminal was selected as the research object and the energy utilization was tested. The voltage, current, imbalance, harmonic, power and energy efficiency of portal crane were analyzed in his paper. Some suggestions were put forward to improve the energy utilization efficiency and reduce energy consumption of portal crane.
Keywords: Port; Portal crane; Energy utilization efficiency; Harmonic.

1 Introduction

Portal crane is one of the main equipment for port shore loading and unloading[1], and also one of the main energy consumption equipment for port production. Therefore, improving the energy utilization efficiency of portal crane through technical and management means plays a significant role in reducing the production energy consumption of bulk terminal. The research on crane energy consumption mainly focuses on the detection of its working state and system. Haibin Yang[2] designed the crane state monitoring system. Hongwei Liu[3] remotely monitored the crane operation state information through different strategies. Bin Chen[4] designed the data warehouse system as the platform to analyze the performance trend of the crane. According to the requirements of "The method of checking and measuring energy utilization efficiency for port electric-crane"[5], this paper monitors and analyzes the energy consumption of portal crane during operation, finds out the shortcomings in energy utilization, and puts forward targeted improvement schemes, which plays an important role in reducing port operation energy consumption.

2 Design of energy efficiency test scheme

2.1 Testing standards and methods

*Corresponding author: eca@foxmail.com

The energy efficiency test shall be carried out in accordance with the requirements of "The method of checking and measuring energy utilization efficiency for port electric-crane" (JTT 314-2009).

The test steps are as follows:

(1) Lift the load at the rated speed. The load mass is 80% of the rated lifting capacity. The amplitude is 50% of the maximum amplitude. And the load lifting height is 10m. Place the load to the lifting position.

(2) The load mass is 80% of the rated lifting capacity. The rotation radius is 50% of the maximum load. Rotate 180 ° clockwise. After the load is stable, rotate 180 ° counterclockwise.

(3) The load mass is 80% of the rated lifting capacity. The amplitude changes from 50% of the maximum amplitude to the maximum amplitude. After the load is stable, return to the position before luffing.

(4) The lifting mechanism rises and falls by 10m at the rated speed, and the amplitude is 50% of the maximum amplitude.

2.2 Testing process

No-load lifting →No-load luffing (smaller) →No-load rotation →No-load luffing (larger) →No-load lowering →Grasp the load →Loaded lifting →Loaded luffing (smaller) →Loaded rotation →Loaded luffing (larger) →Loaded lowering →Load lowering →No-load lifting →No-load luffing (smaller) →No-load rotation → No-load luffing (larger).

3 Voltage and current test and analysis

In this measurement, several working cycles are measured, and the extreme values of voltage and current are shown in Table 1.

Table 1. Voltage and current extremum.

No.	Test content	L1	L2	L3	Average value
1	Maximun voltage(V)	411.45	413.1	411.65	412.1
2	Minimum voltage(V)	356.65	358.15	360.25	358.35
3	Maximum current(A)	572	570	578	573.3
4	Minimum current(A)	7	1	7	5

In this workflow, during heavy load lifting, the voltage decreases, the current increases greatly, the peak value reaches more than 560A, the power factor is 0.99, and the electric energy is converted into gravitational potential energy and kinetic energy. During the intermittent period of loading, the voltage is stable, the current value is the smallest, and it is basically standby loss. When the heavy load drops, the gravitational potential energy is released and converted into electric energy and kinetic energy. The whole process is in a cycle, the current changes periodically, and the voltage decreases when the current increases greatly. At this time, the voltage drop is mainly caused by heavy load startup. The maximum power supply voltage is 413.1V and the minimum voltage is 356.65V. The power supply voltage of portal crane is 380V. The voltage deviation is 6.14% - 8.71%, and 95% of the voltage is distributed within 408.4V, as shown in Figure 1. 95% of the current is distributed within 309 A, as shown in Figure 2.

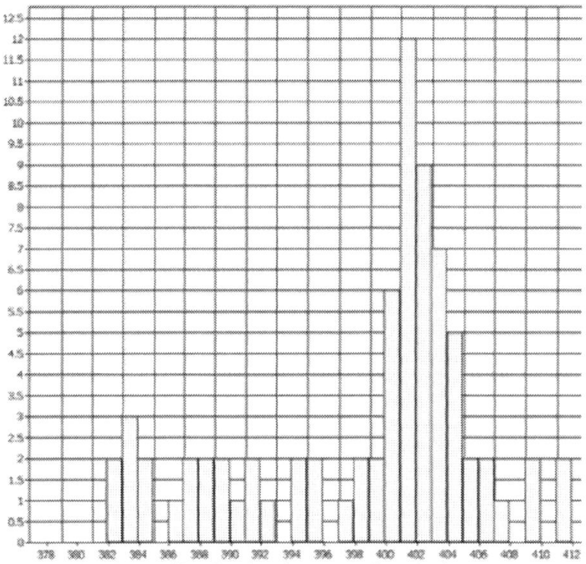

Fig. 1. Probability diagram of voltage distribution in normal period.

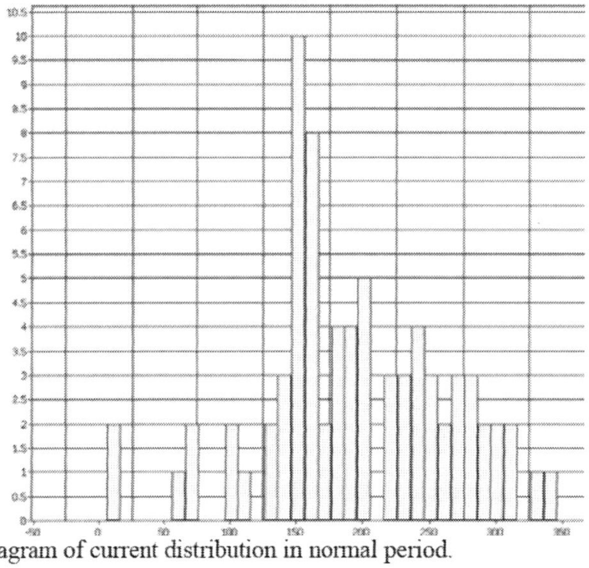

Fig. 2. Probability diagram of current distribution in normal period.

4 Imbalance test and analysis

The imbalance of supply voltage is within 0.65%, less than 2% of the standard value. The current imbalance is basically kept within 15%. The situation of large imbalance value is mainly that the deceleration time and current drop after the lifting weight reaches the required height, and the current difference is small, but when it accounts for a large proportion of the current value, the current imbalance is not more than 5% in the main lifting and rotation process.

The included angle of voltage and current of each phase changes greatly, basically between 0-60 degrees. The reactive power fluctuates greatly, the power factor changes greatly, and the electric energy loss caused by reactive power is large. Especially in the case of no-load, rotating forward and descending with load, according to the working mechanism, except that the power factor of main load lifting is very high, which is between

0.97-0.99, the power factor is between 0.44-0.85 at other times, which is negative when falling. In standby, the power factor is about 0.45.

95% of voltage imbalance is within 0.3%, and 95% of current imbalance is within 15%. Voltage imbalance is better than current imbalance. The possible causes of voltage imbalance are high-order harmonic and reactive power, but the proportion is small and does not affect the actual operation.

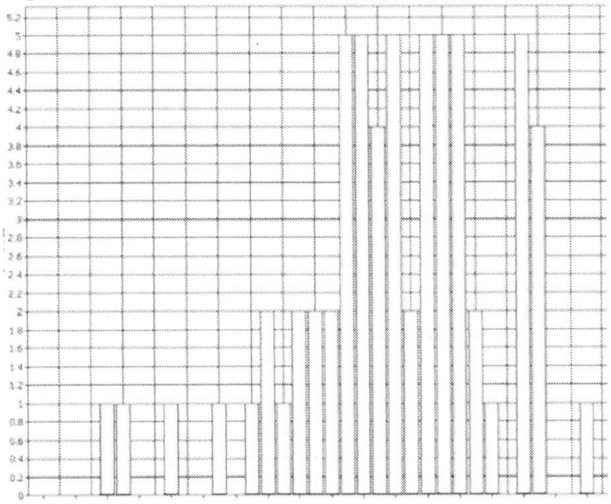

Fig.3. Voltage imbalance distribution.

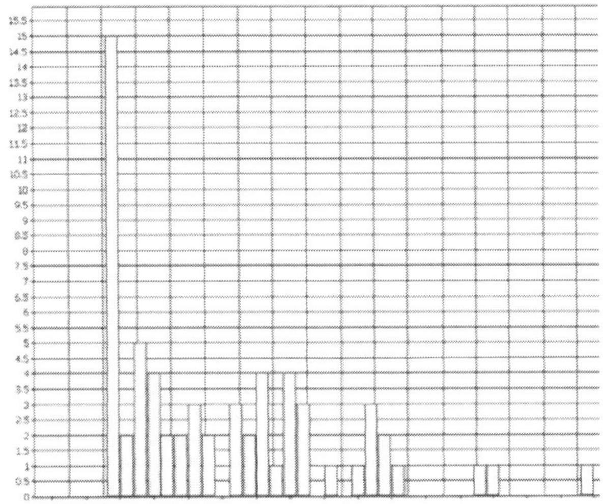

Fig.4. Current imbalance distribution.

5 Power test analysis

The portal crane works periodically. During heavy load lifting, the power consumption is the largest, and the instantaneous power is large. During heavy load falling, the gravity potential energy is mainly in the power generation state, and the instantaneous power is negative.

It can be found that there are positive work and negative work (feedback electric energy) in portal crane. Because the voltage and frequency of feedback electric energy do not match with the power grid, it will not only affect the measurement of electric energy, but also worsen the power quality of the power grid. At the same time, due to the decline of power quality during the operation of portal crane, the feedback power quality is not high, which affects itself or other electrical equipment, such as reducing the service life of lighting equipment. If the feedback electric energy can be effectively collected or reasonably utilized, it will save a lot of electric energy and be conducive to the use of equipment.

The frequency and voltage imbalance are acceptable, while the average power factor is only 0.31. If negative power is not considered, the average power factor is about 0.7, which is relatively low in multiple working cycles. The power factor is high only when it is raised, but the weight in the whole working cycle is low, so the average value is also low.

In a working cycle, the power consumption of bulk cargo handling is basically maintained at 2-3kWh. The loading and unloading are discontinuous. The longer the intermittent period is, the longer the standby time is. The greater the percentage of standby power consumption in each working cycle is, and the more power consumption per ton of cargo is.In this workflow, the lifting operation is the main power consumption stage, with high power factor and fixed power consumption. Other working links have low power factor and large line loss. Slow operation speed will increase the overall power consumption of the working cycle, so appropriately increasing the working speed other than lifting will help to reduce the loss.

6 Conclusions and recommendations

(1) The portal crane generally operates discontinuously. Reducing standby time and increasing work compactness are helpful to improve work efficiency and reduce energy consumption. It is suggested to mainly adopt optimal scheduling to make the process connection closer and reduce the equipment empty consumption in priority to the technical transformation.

(2) When the equipment works for a long time, the average power factor is not high. It can be considered to add dynamic reactive power compensation device in the nearby power transformation and distribution room to reduce reactive power loss and improve power factor.

(3) The working condition of the bulk cargo terminal is poor. It is recommended to regularly remove dust, fixing screws and other maintenance work of the electric control equipment to reduce the failure rate of the equipment and improve the continuous working time of the equipment while increasing the stability of the equipment.

Acknowledgment

Fund Project, Central-level nonprofit scientific research institutes basic research project special funds TKS20200205.

References

1. Xueyun Li, Jun Li. Analysis of on-line monitoring method for port gantry crane [J]. Shanghai Machine Tool, 2017, (1):11-13.
2. Haibin Yang. Design and implementation of intelligent terminal for remote condition monitoring system of large crawler crane [D]. Shanghai Jiaotong University, 2009
3. Hongwei Liu. Research and development of remote monitoring system for special crane in shipyard [D]. Tongji University, 2010
4. Bin Chen. Research and implementation of data optimization of large crane remote monitoring system [D]. Shanghai Jiaotong University, 2009
5. JTT 314-2009. The method of checking and measuring energy utilization efficiency for port electric-crane [S].Beijing: China Communications Press, 2010

Strength analysis of the lightweight-designed power battery boxes in electric vehicle

Yuhang An[#1], *Xishu* Wang[#1], *Nannan* Dou[#1], and *Zhengbin* Wu[*1]

[1] Tianjin Chinese Academy of Sciences Institute of Advanced Technology, No.3 Haitai-Development Six Road, Xiqing District, Tianjin, PR China
[#] These authors contributed equally to this work.

Abstract. The weight lightening of electric vehicles' power battery box not only improves the vehicles' mileage but also extends the service life of the power battery, while in practice the design of lightweight shall not sacrifice its strength for safety reason. Herein, three materials including the glass fiber composite, carbon fiber composite, and metal steel are used to develop the same power battery box and their strengths are simulated. The results show that the sheet molding compound (SMC, a kind of glass fiber composite) power battery pack shall be damaged when the loaded extrusion force is increased to 98.2 kN; for the metal power battery pack, a force of 81 kN would make the deformation more than 30%; while for the carbon fiber composite one, it can stand the extrusion test up to 100 kN, and the specific gravity and specific strength are higher than the former two materials under the same design model condition. This study proves the carbon fiber composite has the advantages of high strength and light weight when applied to the power battery box.
Keywords: Carbon fiber composites; Simulation analysis; Power battery box; Lightweight; Specific strength.

1 Introduction

With the rapid growth of the number of vehicles during the last decade, new energy vehicles are emerging as the next generation transportation to alleviate the strong dependence of the automobile industry on fossil oils and to reduce pollutant emissions. The development of new energy electric cars, as a strategic developing sector, shall thus optimize the energy structure and protect the natural environment, which is critical to the overall as well as long-term social and economic development.[1] As one of the core technologies of new energy electric vehicles, power battery box has exerted great influence on the car's performance. The pursuit of lightweight power battery box is urgently needed before the substantial scientific advances in batteries, motors, electronic controls, etc.[2]

According to previous studies, it is highly probable to find a high-strength, light-weight, and low-cost box material for an electric car power battery.[3] In particular, the lightweight design of power battery box materials needs not only to reduce the weight, but also to fulfill the necessary criteria of vehicle strength and safety. For example, the ultra-light concept

[*] Corresponding author: zb.wu@siat.ac.cn

automobile was created by General Motors in the United States as early as 1992. The car's body was made of carbon fiber composite materials and weighed 191 kg in total.[4] Sunarto Kaleg et al.[5] improved the design based on the optimized material thickness and achieved the highest quality battery case by using 5052-0 series aluminium alloy as the battery materials. Despite the fact that significant progresses have been accomplished in this field, few studies were performed in regard to lightweight materials comparison for power battery packs under extrusion conditions. A detailed analysis would be helpful and in return advance this field.

In this work, the most used materials including sheet molding compound (SMC, a kind of glass fiber composite), carbon fiber reinforced plastics (CFRP), and metal are selected and modelled to study the effect of extrusion force. To simplify some non-structural parts in power battery packs, 3D modelling software and finite element pre-processing software are applied to mesh the power battery pack of electric vehicles. The properties of materials and boundary conditions of the simulated battery pack are created based on the actual situation.[6] Based on simulation results, the strength of battery packs made of three materials are compared under the same standard operating conditions, and the optimal option for lightweight power battery packs is thereafter obtained.

2 Materials and experiments

2.1 Materials and modelling

This study is based on a company's mini pure electric car, which has met legal criteria but needs further modification to lower energy consumption due to the overweighted power battery box. To achieve this, the lightweight materials of power battery boxes, i.e., glass fiber reinforced plastics (GFRP) composed of thermosetting material sheet molding compound (SMC), carbon fiber reinforced plastics (CFRP), and metal material Q235 steel are chosen as the research objects in this study. Notably, SMC is chosen since it is primarily used in the molding industry as a prepreg of GFRP.[7] It is also widely utilized in automobile body panels, high-strength electrical components, and numerous structural sections. In term of strength, elastic isotropic model with strength criterion on maximum equivalent stresses is chosen and shall be enough for evaluation.

The parameters for each material are set as following. (1) For SMC material, data obtained from actual static compression tests and other similar studies by authors are used as model parameters and ultimate stress, as listed in Table 1. (2) For CFRP material, it is designed to be made up of epoxy resin (or other resins) and carbon fiber according to conventions. Carbon prepreg's strength is directional because of the anisotropic nature of carbon fiber. The mixture of twill cloth and unidirectional fiber is created to suit the strength criteria. Parameters for woven and unidirectional monolayers, obtained from experiments, are presented in Table 2. A bilinear isotropic plastic hardening model with elastic material is used in simulation. (3) For Q235 steel material, wide-accepted characteristics are employed, as shown in Table 3.

In line with the aim of this study and the actual situation of the structure, the pre-processing software are used to simplify and mesh the battery pack structure, and the finite element model of the battery pack are established. The finite element model of the battery pack is dominated by quadrilateral shell elements, supplemented by a small amount of triangle elements, and extrusion cylinder is set as a solid element.

 https://doi.org/10.1051/e3sconf/202234101025

Table 1. Physical and mechanical properties of SMC composite.

Parameter	Density (kg/m3)	E (GPa)	Poisson's ratio μ	σcrit (MPa)
SMC	2251	12.5	0.3	120

Table 2. Physical and mechanical properties of CFRP materials.

Parameter	Woven fabric	Unidirectional tape T700
E1 (GPa)	159.520	245
E2 (GPa)	11.660	6.9
G12 (GPa)	3.813	3.9
G13 (GPa)	3.813	3.9
G23 (GPa)	3.961	3.0
μ12	0.267	0.87
σв1 + (MPa)	2005	2160
σв2+ (MPa)	68	45
σв1 - (MPa)	1355	980
σв2- (MPa)	198	140
Interlaminar shear strength τ13 (MPa)	60	83
Monolayer thickness (mm)	0.3	0.3

Table 3. Physical and mechanical properties of Q235.

Parameter	Density (kg/m3)	Yield stress σT (MPa)	Ultimate stress σв (MPa)	Poisson's ratio μ	Young modulus E (GPa)
Q235	7.85	235	400	0.3	200

2.2 Extrusion experiment design

The experiments in this study are carried out following the condition and programme requirements of the Chinese national standard "GB/T 38031-2020 safety requirements standard for power battery for electric car" for power battery box, as listed below in details.

1) The radius of the extrusion cylinder shall be 75 mm, and the height of it shall be more than the test object but no more than 1 m.
2) The extrusions shall be carried out along X and Y direction, which represent the driving direction of the vehicle and the perpendicular direction of the driving direction in the horizontal plane, respectively.
3) When the extrusion force exceeds 100 kN or the extrusion distortion reaches 30% along the extrusion direction, the extrusion shall be stopped.
4) The extrusion speed shall be no more than 2 mm/s.

 https://doi.org/10.1051/e3sconf/202234101025

3 Results and discussion

3.1 Structural design and strength calculation analysis of SMC power battery box

In simulation, a model of elastic isotropic material with strength criteria based on maximum equivalent stress and parameters from Table 1 is used. When equivalent stress reaches ultimate value, it is removed, and then the redistribution of stress occurs. Fig. 1(a) and (b) shows stress-strain state of a battery pack box parts made of SMC during fracture.

At the same time, the model allows estimation of damage evolution in composite and its influence on bearing capacity. Parts of the box are connected and have contact interaction with experimental tool's cylinders. Boundary conditions were changed for the definition of a bearing capacity. Displacement of one of the cylinders is limited at all degrees of freedom, while the second cylinder can only be displaced along the battery pack box on a distance corresponding to maximum permissible compression. In such conditions, the maximum stress results in bearing capacity failure and is equal to F_{SMC} = 98.2 kN.

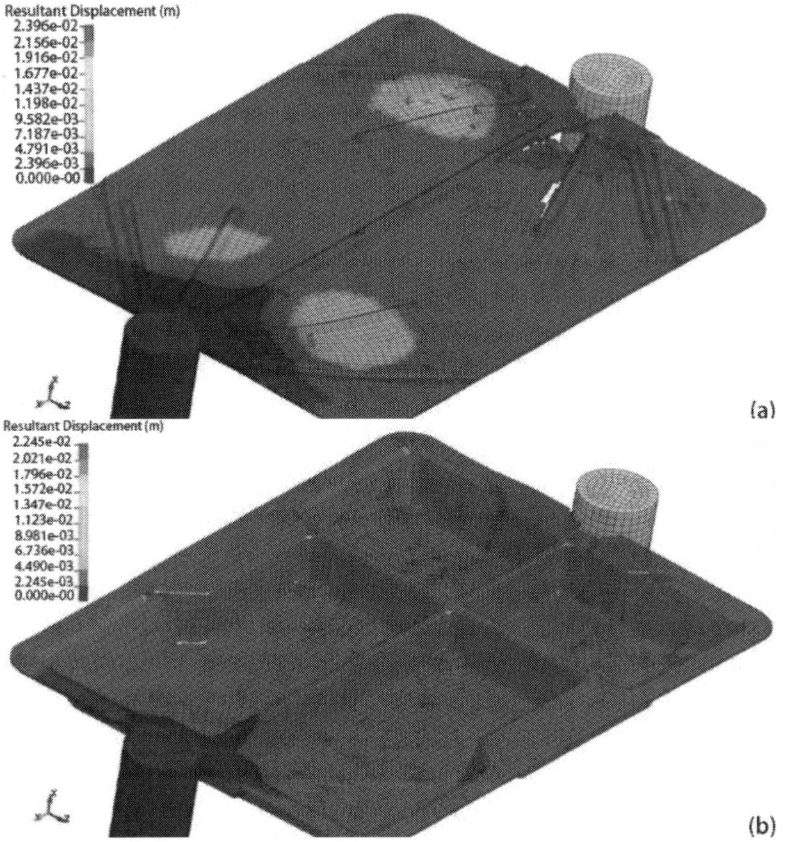

Fig. 1. Displacement analysis of SMC power battery box during fracture. (a) Upper part, (b) Lower part.

 https://doi.org/10.1051/e3sconf/202234101025

3.2 Structural design and strength calculation analysis of CFRP power battery box

Similar to power battery box made of SMC, a design of a battery pack box with maximum bearing capacity was developed with the employment of high anisotropic material - CFRP. In this study, carbon fabric consists of woven and unidirectional layers. A model of a multi-layered composite material in quasi-static setting is used within the borders of a plane stress state.

A finite element model of a composite box was calculated. Building of a geometrical model with given parameters occurs in "Geometry" module. Parameters of braided monolayer and unidirectional monolayer obtained in laboratory are shown in Table 2. Fig. 2(a) illustrates the visualization of zones thickness in accordance with complex layup scheme. Surfaces are divided into many fabric and unidirectional tape layup zones for maximum bearing capacity. Fig. 2(b) shows CFRP box deformation in axial direction. According to the extrusion test method, the carbon fiber composite power battery box does not produce damage until the maximum force reaches F_{CFRP} = 115.6 kN.

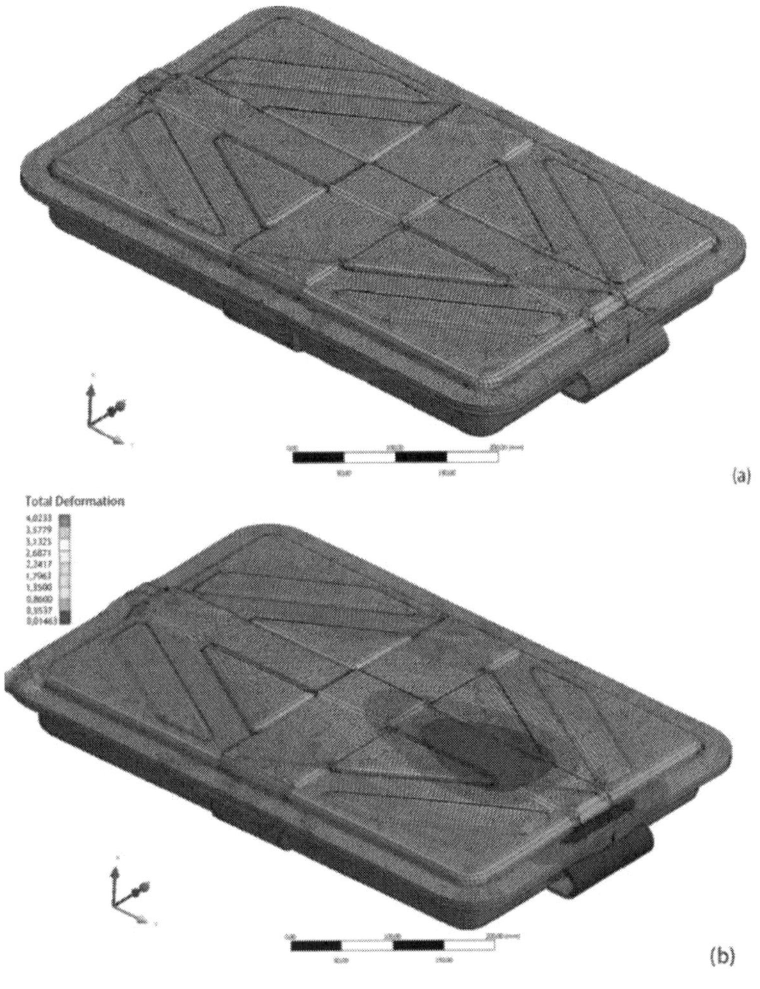

Fig. 2. Analysis of CFRP power battery box. (a) Finite element model, (b) Box deformation in axial direction.

https://doi.org/10.1051/e3sconf/202234101025

3.3 Structural design and strength calculation analysis of metal power battery box

The Q235 steel is employed in the design of steel power battery boxes and its dynamic strength is simulated. Fig. 3(a) and (b) depict cloud diagrams of plastic strain distribution. It is seen that there is a substantial plastic strain in the steel power battery box when the cylinder of the experimental tool is hidden, but the stress level is lower than the maximum strength. The maximum transverse compression is about 30% when the maximum extrusion force reaches F_{Q235} = 81.0 kN.

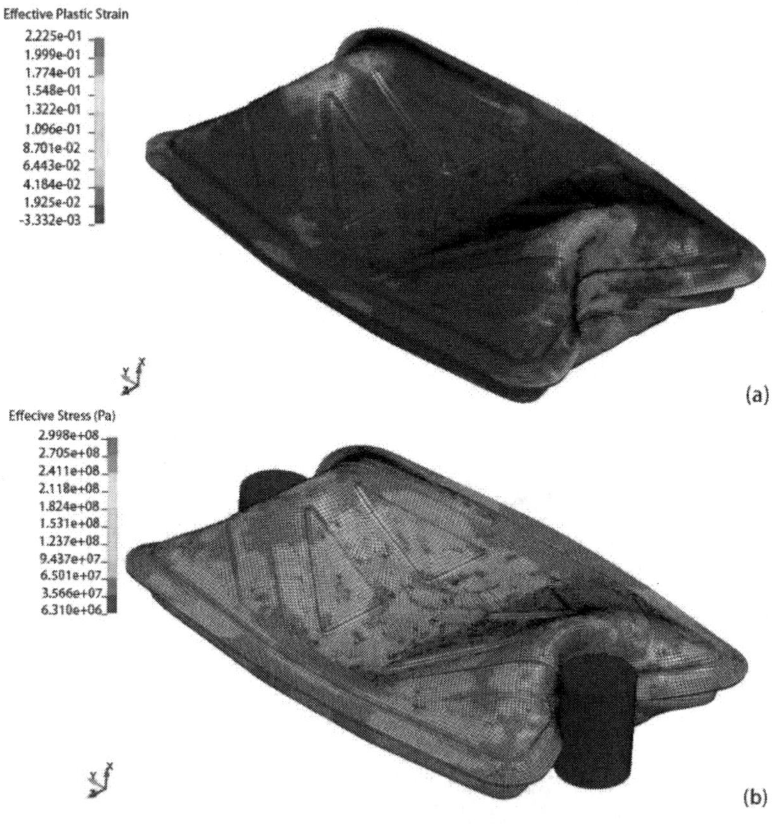

Fig. 3. Analysis of steel power battery box. (a) Effective plastic strain, (b) Effective stress.

3.4 Comparison of structural design results of power battery box of different materials

Table 4 tabulates figures of merit for power battery boxes made of various materials, as analysed above. The simulations reveal that the CFRP power battery box has the maximum specific strength, while the steel power battery box has the lowest. Remarkably, the specific strength of CFRP power battery box is 2.58 times that of steel one.

In order to conduct a more precise comparison of characteristics of power battery box made of different materials, a specific parameter is introduced. For estimations of strength, a maximum load F that leads to the loss of a bearing capacity during axial compression is defined from simulation experiments. A specific strength is defined as the ratio of maximum load to power battery box mass, i.e.,

$$\sigma = \frac{F}{M} \quad (1)$$

Specific strength allows preliminary evaluation of strength efficiency of a whole class of promising designs without engineering calculations.

Table 4. Physical and mechanical properties of Q235.

Material	Weight (g)	Maximum-loaded force (kN)	Specific strength, σ
SMC	9110	98.2	10.7
Metal	10470	81.2	7.7
CFRP – Tape+Fabric	5784	115.6	19.9

4 Conclusion

In summary, three different types of materials including SMC, CFRP, and metal were used to design power battery box in electric vehicles, and their performance are assessed according to simulations. The extrusion simulation showed that to meet the strength requirements of related electric vehicle power battery boxes, the layer-structured CFRP material with unidirectional belt and fabric had the most significant weight reduction effect, followed by GFRP and then metal materials. Compared with the metal-made power battery box, the CFRP material battery pack could reduce the weight by 44.7%, while enhance the specific strength by 158.4%. This study enriches the current theoretical research of the power battery box, and offers more opportunities and effective ways to guide the further lightweight design of the new energy electric vehicles in the aim of meeting the requirements of related extrusion processes.

Acknowledgement

This work is supported by Tianjin "The Belt and Road Initiative" technical R & D cooperation and industrialization projects (19YDYGHZ00070).

References

1. S. Chen, Y. Feng, C. Lin, Z. Liao, Mei, X. Math. Probl. Eng. **2021**, 6613602 (2021)
2. H. Zhou, W. Feng, L. Sun, J. Asian Elect.Vehicles **2**(1), 531-534 (2009)
3. M. Hartmann, M. Roschitz, Z. Khalil, Mater. Sci. Forum **765**, 5 (2013)
4. H. Fuchs, USA: 2007 SPE Automstive Composites Conference, 2-17 (2008)
5. S. Kaleg, Amin, 2016 International Conference on Sustainable Energy Engineering and Application (ICSEEA), pp. 1-5, doi: 10.1109/ICSEEA.2016.7873558 (2016)
6. X. Chen, X. Sun, P. Chen, B. Wang, Y. Zhao, Compos. Struct. **256**, 113120 (2020)
7. V. Feuillade, A. Bergeret, J.C. Quantin, A. Crespy, Compos. Sci. Tech. **66**(1), 115-127 (2006)

Techno-economic analysis of 30MW biomass power plant modification for heating supply

Wei Qian[1], *Jiawei* Qiao[2], *Bin* Li[2], *Xing* Zheng[2], *Zongming* Zheng[2], and *Xianbin* Xiao[2,*]

[1]SPIC Capital Holding Co., Ltd, Beijing 100045, China
[2] National Engineering Laboratory for Biomass Power Generation equipment, North China Electric Power University, Beijing 102206, China

Abstract. Biomass combined heat and power (CHP) generation is the more efficient use of biomass compared the direct combustion power generation. The techno-economic analysis of 30 MW biomass direct combustion power set's modification for heat supply is investigated in this work. Three schemes are employed to provide the district heating supply, including low vacuum condenser, steam extraction, and turbine bypass. The low vacuum condenser approach is the most promising based on the techno-economic analysis. The calculated outlet pressure of the steam turbine, the power generation capacity, the heating supply and the maximum heating area are 30KPa, 27MW, 143.58GJ/h, and 900,000 m^2 respectively. The expected annual income of biomass combined heat and power plants amounts up to 7,422,700 RMB.
Keywords: Biomass combined heat and power, Low vacuum; Steam extraction, Bypass heating.

1 Introduction

Energy transition towards a low-carbon and high-efficiency energy system has been a long-term strategy for China. Biomass plays an important role in energy mix[1]. Biomass power plants could provide heating and electricity to meet the heating requirement, which is the world's largest thermal power market [2]. Biomass power plant modification for heating supply is an important approach to reduce energy consumption, increase energy supply, and improve energy efficiency.

Some efforts of heating modification have been made in coal-fired boilers of thermal power plants. Xu et al. adopted the steam supply method of reheating the cold section of the steam turbine and the exhaust steam extraction of the medium pressure cylinder and configuring the pressure matcher[3]. Liu et al. proposed to reduce the work share of the high-pressure cylinder, so as to increase the main steam flow of the unit and the heating and pumping capacity of the unit under the same power generation power conditions[4]. However, the modification of small capacity biomass power plant needs the techno-economic analysis in details.

*Corresponding author: xiaoxianbin@ncepu.edu.cn

2 Case description

A biomass power plant has a 30MW generator set, steam turbine model is N30-8.83 with high temperature, high pressure, single cylinder, single shaft, condensing type, six-stage reheat extraction. The recuperation system consists of 2 high-pressure heaters, 1 shaft seal heater, 1 deaerator, 3 low-pressure heaters and low-heating trap pumps[5]. The main parameters of the unit are shown in Table 1, and the pumping parameters at all levels of the recuperation system are shown in Table 2.

Table 1. Turbine design parameters.

Parameter	Unit	Numerical value
Rated power	MW	30
Rated steam inlet	t/h	114
Rated exhaust pressure	kPa	4.9
Cooling water temperature	℃	20
Condenser steam flow	t/h	79.5
Cooling water volume	t/h	4500 ~ 6500
Rated inlet steam temperature and variation range	℃	535 (+5/-10)
Rated steam inlet pressure and variation range	MPa	8.83±0.49

Table 2. Steam extraction at all levels in full load operation.

Parameter	Pressure (MPa)	Temperature (℃)	Flow(t/h)
After adjustment	4.608	459	
The first level (to #1 high plus)	2.93	405	8.909
Second level (to #2 high plus)	1.143	297	1.4
Third pole (to deaerator)	0.649	239	4.374
Level 4 (to #4 low plus)	0.344	180	4.5
Level 5 (to #5 low plus)	0.165	119	5.209
Sixth level (to #6 low plus)	0.063	87	8.573

3 Modification schemes

3.1 Low vacuum circulating water heating

The principle of this technology is to reduce the vacuum degree of the steam turbine condenser, increase the exhaust steam temperature of the steam turbine, and then increase the return water temperature of the cooling water of the condenser for heating[6]. The circulating water of the heating network passes through the condenser for the first heating and absorb the waste heat of exhaust steam. In turn, the heating network heater is used for secondary heating to generate high-temperature hot water[7]. Then the generated hot water is sent to the hot water pipe network, through the heat exchange station and circulating water for heat exchange. The process is shown in Fig. 1.[8].

The generating power of the unit after the transformation:

$$P_1=(h_1-h_2)\cdot D_t+(h_2-h_3)\cdot(D_t-D_1)+(h_3-h_4)\cdot(D_t-\sum_{i=1}^{2}D_i)+(h_4-h_5)\cdot(D_t-\sum_{i=1}^{3}D_i)+(h_5-h_6)\cdot(D_t-\sum_{i=1}^{4}D_i)+(h_6-h_7)\cdot(D_t-\sum_{i=1}^{5}D_i) \quad (1)$$

Where h_1 is the steam turbine inlet steam enthalpy, kJ/kg; h_2-h_6 is Steam enthalpy of each level of extraction steam, kJ/kg; h_7 is the enthalpy of outlet steam, kJ/kg; D_t is the main steam flow, kg/h; D_1-D_5 is the extraction steam volume at each stage, t/h.

Heat energy produced:

$$Q_1=(h_7-h_{njs})\cdot D_m\cdot\eta_1\cdot(1-\eta_2) \quad (2)$$

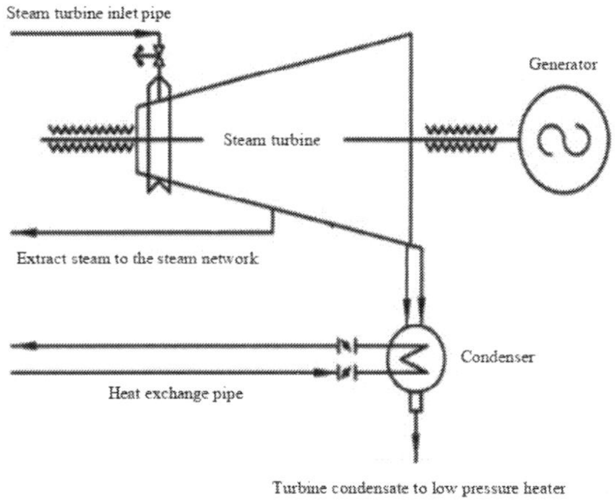

Fig. 1. Schematic diagram of low vacuum circulating water heating.

Where h_{njs} is the enthalpy of the condensate, kJ/kg; D_m is the outlet steam flow, kg/h; η_1 is the efficiency of the heat exchanger, %; η_2 is the pipeline loss, %.

Increased temperature of cooling water:

$$\Delta t = \frac{Q_1}{C_p D_l} \tag{3}$$

Where D_l is the cooling water flow, t/h.

Considering various factors, this paper selects five sets of steam turbine outlet steam pressure to calculate the corresponding power generation and heat supply. The heating heat index is usually 60W/m^2. The calculation results are presented in table 3.

Table 3. Low vacuum circulating water transformation calculation result.

Parameter	Unit	Heating period					Non-heating period
Outlet pressure	kPa	35	30	25	20	15	4.9
inlet enthalpy	kJ/kg	3476.67	3476.67	3476.67	3476.67	3476.67	3476.67
Exit enthalpy	kJ/kg	2307.93	2287.08	2262.85	2233.81	2197.35	2065.03
Power generation	KW	26839.32	27268.72	27767.67	28365.80	29116.68	29294.42
Heat supply	GJ/h	145.07	143.58	141.84	139.77	137.16	——
Heating area	× 10^{4}m^2	1.362	90.42	89.33	88.022	86.38	——
Return water temperature	℃	61.55	61.43	61.29	61.13	60.92	——

3.2 Steam extraction

Due to the working nature of the deaerator, power system emits a large amount of high-temperature steam, which not only causes energy waste, but also produces certain pollution to the environment[9]. The so-called retrofit of external heating and hot water supply are to extract a certain amount of steam from the holes reserved for the deaerator for heating[109]. The specific process is shown in Fig. 2.

The generating power of the unit after the transformation:

$$P_2 = (h_1 - h_2)\cdot D_t + (h_2 - h_3)\cdot(D_t - D_1) + (h_3 - h_4)\cdot(D_t - \sum_{i=1}^{2} D_i) + (h_4 - h_5)\cdot(D_t - \sum_{i=1}^{3} D_i - D_a) + (h_5 - h_6)\cdot(D_t - \sum_{i=1}^{4} D_i) + (h_6 - h_7)\cdot(D_t - \sum_{i=1}^{5} D_i) \tag{4}$$

Where D_a is the flow of additional steam extracted from the deaerator, t/h.

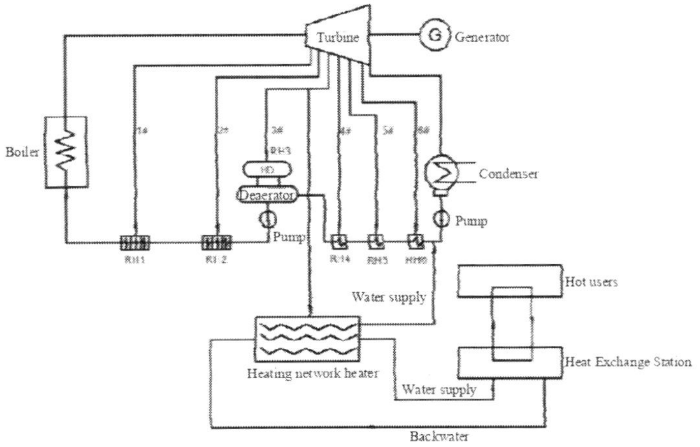

Fig. 2. Schematic diagram of Steam extraction.

Heat energy produced[11]:

$$Q_2 = D_a(h_5 - h_{5a}) \cdot \eta_1 \cdot (1-\eta_2) \tag{5}$$

Where h_{5a} is the steam extraction pressure corresponding to the saturated water enthalpy, kJ/kg;

Various factors are comprehensively considered, and five groups of different extraction steam quantity are selected to calculate the power generation and heat supply. The results are shown in Table 4.

Table 4. Steam extraction retrofit calculation result.

Parameter	Unit	Heating period				
Steam extraction	t/h	55	50	45	40	35
Steam extraction enthalpy	kJ/kg	2932.61	2932.61	2932.61	2932.61	2932.61
Extraction steam pressure corresponds to saturated water enthalpy	kJ/kg	683.99	683.99	683.99	683.99	683.99
Power generation	KW	27703.06	28124.49	28241.48	28358.47	28475.47
Heat supply	GJ/h	111.31	101.19	91.07	80.95	70.83

3.3 Bypass heating retrofit

The heat load is certain when the heat load is fixed, so the unit cannot achieve true thermolytic coupling in the above two cases. Many power plants can only use one unit with heating and the other with peak shaving. Some scholars have proposed the solution to the boiler with heating, that is, the heating load is completely carried by the boiler, and the heating steam and power generation steam are separated at the furnace end, and no longer enters the steam turbine to perform work[12]. The specific process is shown in Fig. 3.

The generating power of the unit after the transformation:

$$P_3 = (h_1 - h_2)\cdot D_t' + (h_2 - h_3)\cdot(D_t' - D_1) + (h_3 - h_4)\cdot(D_t' - \sum_{i=1}^{2} D_i) + (h_4 - h_5)\cdot(D_t' - \sum_{i=1}^{3} D_i) + (h_5 - h_6)\cdot(D_t' - \sum_{i=1}^{4} D_i) + (h_6 - h_7)\cdot(D_t' - \sum_{i=1}^{5} D_i) \tag{6}$$

Where D_t' is the steam flow used for power generation, t/h.

Heat energy produced:

$$Q_3 = (D_t - D_t')\cdot h_1 \cdot \eta_1 \cdot (1-\eta_2) \tag{7}$$

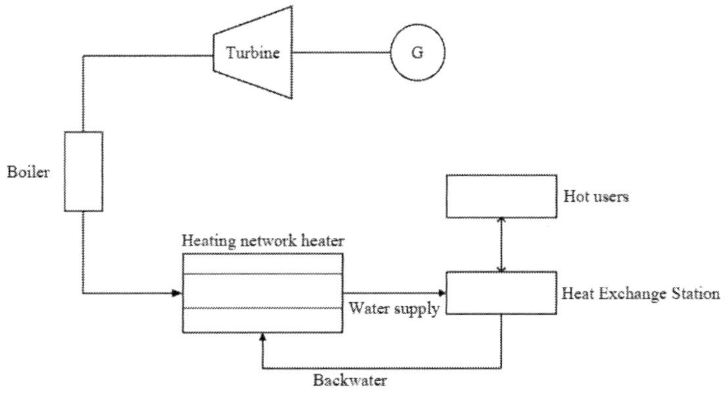

Fig. 3. Schematic diagram of bypass modification.

Various factors are comprehensively considered, and five groups of different extraction steam quantity are selected to calculate the power generation and heat supply. The results are shown in Table 5.

Table 5. Bypass reconstruction calculation results.

Parameter	Unit	Heating period				
Steam extraction	t/h	20	25	30	35	35
Power generation	KW	28294.12	27198.21	26094.85	24775.64	22987.34
Heat supply	GJ/h	69.5334	86.91675	104.3001	121.6835	139.0668
Heating area	$10{,}000m^2$	39.41164	49.26455	59.11747	68.97038	78.82329

4 Economic analysis

4.1 Profit analysis

In the heating season of each year, the unit runs at full load for 3000 hours, and the heat supply during the heating period is always. Calculated based on heat price of 31.5 yuan/GJ[13], the revenue from the co-sale of heat supply during the heating period each year is:

$$Y_1 = 31.5 \times 3000Q \tag{8}$$

Where Q is the steam heat used for heating in the three scheme, kJ/kg. Y is the specific income.

4.2 Power loss

Calculated based on the biomass feed-in subsidy price of 0.75 yuan/kWh[14], the heating season (3000h) will reduce the income from external power supply:

$$Y_2 = 0.75 \times 3000\Delta P \tag{9}$$

Where ΔP is the power generation loss due to heating renovation.

4.3 Net income

From the above, we can know net income is:

$$Y = Y_1 - Y_2 \tag{10}$$

The economic calculation results of the three schemes are shown in Table 6.

Table 6. Economic calculation results of the transformation plan.

	Parameter	Unit	Heating period				
Low vacuum circulating water heating reform	Heat supply	GJ/h	145.07	143.58	141.84	139.77	137.16
	Heating revenue	10,000RMB	1370.91	1356.81	1240.43	1220.79	1196.14
	Retrofit loss	10,000RMB	711.15	614.54	602.27	567.69	498.74
	Net income	10,000RMB	659.76	742.27	638.16	653.10	697.39
heat exchange external heating hot water heating transformation technology	Heat supply	GJ/h	111.31	101.19	91.07	80.95	70.83
	Heating revenue	10,000RMB	1051.85	956.23	860.60	764.98	669.36
	Retrofit loss	10,000RMB	516.81	421.99	395.67	369.34	343.02
	Net income	10,000RMB	535.04	534.24	464.94	395.64	326.34
Bypass heating retrofit	Heat supply	GJ/h	69.53	86.92	104.30	121.68	139.07
	Heating revenue	10,000RMB	657.09	821.36	985.64	1149.91	1314.18
	Retrofit loss	10,000RMB	383.82	630.40	878.66	1175.48	1577.85
	Net income	10,000RMB	273.27	190.96	106.98	-25.57	-263.67

It can be seen from the table that when the low-vacuum circulating water heating technology is adopted and the outlet pressure of the steam turbine is 30KPa, the maximum net income is RMB 7,422,700. Adopting heat exchange external heating and hot water renovation technology, when the air extraction volume from the deaerator is 55t/h, the maximum net profit is 5,350,400 yuan. By adopting the bypass heating reform technology, when the pumping capacity is 20t/h, the maximum net profit is RMB 2,732,700.

5 Conclusion

When the low vacuum circulating water heating technology is adopted, it is the optimal solution to increase the outlet steam pressure to 30kPa. At this time, the heat supply is 143.58GJ/h, the maximum heating area reaches about 900,000 m^2, and the annual income is 7,422,700 Yuan. When adopting the heat exchange external heating hot water transformation technology, the pumping steam volume is equal to 55t/h as the optimal solution, the heat supply is 111.31GJ/h, the maximum heating area is about 630,900 m^2, and the annualized income is 5,350,400 Yuan. Using bypass heating transformation technology, the most suitable solution is to enter the steam turbine when the rated steam intake is 20t/h, the heat supply is 69.53GJ/h, the maximum heating area is about 394,100 m^2, and the annualized income is 2,732,700 yuan.

Comprehensive comparison of the three transformation schemes, low vacuum circulating water heating transformation technology is more mature, under the premise of meeting the overall performance of the unit, it can reduce the overall energy consumption of the unit and ensure the stable operation of the device, and the operating economy is high, which is the best modification scheme.

References

1. J. Li, W. Cai, C Wang, Y Chen. Research on the deployment strategy of biomass

energy technology in China's power sector under the vision of carbon neutrality [J]. China Environmental Management, 13(01): 59-64(2021).

2. S. Chen. Accelerate the sustainable development of agriculture and forestry biomass power generation industry [N]. China Agricultural Mechanization Herald,005(2021).
3. R. Xu, J. Zhao, D. Zhang. Design and control of 600MW unit heating retrofit system[J]. China Electric Power, 48(07): 72-75(2015).
4. C. Liu, L. Geng, H. Wang, H. Miao, Z. Shang, L. Zeng. Research on the control strategy of high and low bypass combined heating reformation[J]. Thermal Power Generation, 49(11): 126-132(2020).
5. Y. Chen, Z. Li, Z. Jiang, X. Chen, Q. Fu, L. Hu. 200MW unit low vacuum heating retrofit plan and energy saving analysis [J]. Steam Turbine Technology, 61(06): 468-470+420(2019).
6. X. Yan. Research on the application of low vacuum circulating water heating technology in thermal power plants [D]. Shenyang Jianzhu University(2019).
7. Y. Yangn. Low vacuum retrofit and economic analysis of condensing steam turbine unit[J]. Large Nitrogen Fertilizer, 43(01): 65-68(2020).
8. J. Liang. Research on integration of a new type of combined heat and power central heating system and optimization of heating network parameters [D]. Yanshan University(2020).
9. Y. Cao, H. Peng, J. Wu, L. Zhang, X. Chen. Cause analysis and treatment of overpressure in the deaerator of ultra-supercritical coal-fired units[J]. Electromechanical Information, (23): 22-23(2020).
10. T. Sun. Evaluation of energy efficiency index of back pressure cogeneration unit[J]. Energy Conservation, 39(09): 49-53,(2020).
11. S. Zheng, K. Chen, G. Fan, K. Chen, C. Gao, Y. Dai. Research on decoupling control characteristics of supercritical extraction steam turbines[J]. Chinese Journal of Power Engineering, 41(03): 201-207+213(2021).
12. L. Wang Lina. Analysis of the relationship between the electric power of the steam turbine unit and the amount of heating steam[J]. Internal Combustion Engines and Accessories, (10): 94-95(2020).
13. X. Xu, L. Tang, C. Zhou, R. Yang. Retrofit of adding a back pressure turbine generator to the industrial heating of a 330 MW coal-fired unit[J]. Power Generation Equipment, 34(06): 434-438(2020).
14. S. Liu, J. Zhang, X. Wu, Y. Hu. Zero output performance analysis and application research of low pressure cylinder of 320 MW unit[J/OL]. China Electric Power: 1-8 (2021).

Method of forest fire automatic early warning and alarm system design

Yucun Wang[1], *Fengyun* Ma[2,*], *and Hao* Fan[1]

[1]Shandong Agricultural University School of Information Science and Engineering, Tai'an City, Shandong Province, China.
[2]Shandong Agricultural University School of forestry, Tai'an City, Shandong Province, China.

Abstract. Forest is a huge green wealth bestowed by nature to mankind, but forest fire will cause great harm. Automatic forest fire early warning and alarm system can warn the high fire risk areas in the forest, or find and alarm the forest fire as early as possible, which is of great significance to reduce the loss of forest fire. This paper introduces a construction method of forest fire automatic early warning and alarm system, which provides knowledge and experience accumulation for the subsequent research and development of related systems.

1 Introduction

Forest is the largest terrestrial ecosystem on earth and an important part of the global biosphere. It is the gene pool, carbon pool, water pool and energy pool on the earth. It plays a vital role in maintaining the ecological balance of the whole earth. It is the resource and environment for human survival and development. Forest is known as the "lung of the earth". Forest is a huge green wealth bestowed by nature. By the end of the Eighth National Forest Resources Inventory, the physical stock of forest resources assets in China was 16.074 billion cubic meters; and the total value of forest land resources assets was 21.29 trillion CNY. The harm of forest fires is amazing. Taking 2016 as an example, China had 2034 forest fires in 2016, with a loss of 60846 cubic meters of forest stock, 137.65 million young trees, and other losses caused by forest fires converted into 41.36 million CNY.[1] Therefore, it is of great significance to build an automatic forest fire early warning and alarm system to warn the high fire risk areas in the forest, or to find and alarm the forest fire as early as possible, which is of great significance to protect forest resources and reduce the loss of forest fire.

2 Overall system structure

The automatic forest fire early warning and alarm system consists of on-site information acquisition subsystem, information transmission subsystem and information processing center subsystem. The system's structure is shown in Figure 1. The input of the system

* Corresponding author: sdmfy@sdau.edu.cn

comes from various sensors of the field data acquisition subsystem, and the output of the system is fire warning signal.

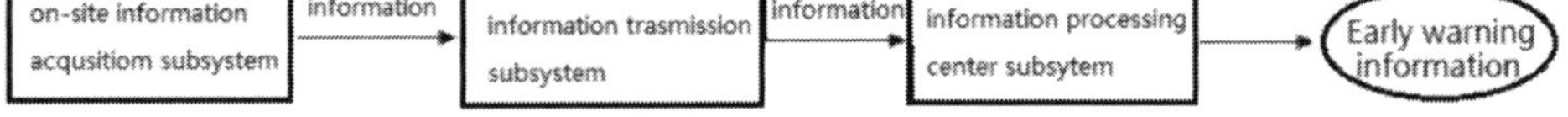

Figure 1. Overall system structure of Forest fire automatic early warning and alarm system

Fig. 1. Overall system structure of Forest fire automatic early warning and alarm system

2.1 Field information acquisition subsystem

The occurrence of forest fire is related to the moisture content, temperature, humidity, oxygen concentration and other factors of combustibles in the forest[2], while the spread of forest fire is related to the types of combustibles, terrain, wind speed and other factors[3][4]. The on-site information acquisition subsystem is the subsystem that completes the relevant information data acquisition for the early warning and alarm system of forest fire. Using the sensor network composed of various sensors, the on-site information acquisition subsystem can collect various relevant data in the forest fire monitoring area in real time, and transmit the data to the information processing center subsystem through the information transmission network, so as to provide data support for the forest fire early warning and alarm of the information processing center subsystem. The system structure of the field information acquisition subsystem is shown in Figure 2.

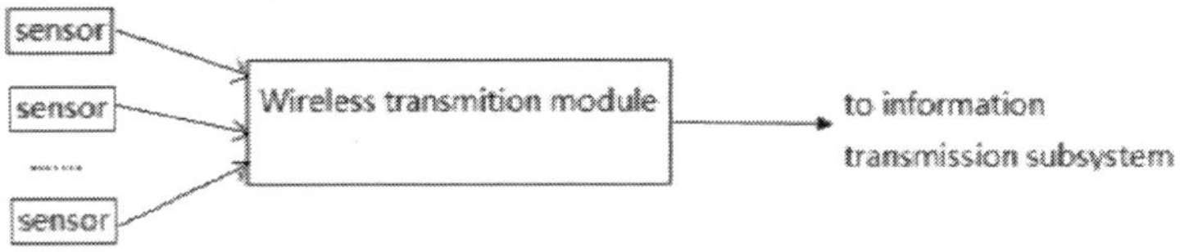

Fig. 2. Field information acquisition subsystem

2.2 Information transmission subsystem

The information transmission subsystem completes the information transmission task in the forest fire automatic early warning and alarm system. Since there is no controllable equipment capable of extinguishing forest fires so far, the system does not consider the control of on-site execution equipment, that is, the system does not design the control information transmission from the subsystem of information processing center to the site, Therefore, the subsystem of forest fire automatic early warning and alarm system is only responsible for transmitting various sensing data collected by the on-site information collection subsystem to the subsystem of information processing center in an appropriate information transmission mode. Considering the influence of various factors, the information transmission subsystem mainly adopts 4G or GPRS to complete the information transmission. For forest areas not fully covered by 4G or GPRS, satellite communication can be considered to complete the function of information transmission subsystem if conditions permit.

2.3 Information processing center subsystem

The information processing center subsystem is the information processing center and decision-making center of the whole system. The information processing center subsystem receives various field data information transmitted by the field information collection subsystem through the information transmission subsystem, analyzes and processes the data through a certain algorithm, studies and judges the forest fire risk level of the forest area where each sensor node is located, and gives the corresponding information early warning. If the subsystem of the information processing center studies and judges the occurrence of forest fire for the information transmitted from the sensor, the subsystem of the information processing center shall issue an emergency fire alarm accordingly, and give specific fire geographic location information and more reasonable forest fire extinguishing plan, so as to provide time conditions for extinguishing forest fire as soon as possible. In order to complete the above functions, the subsystem of the information processing center shall not only receive the sensing information, but also locally maintain the database containing meteorological information, historical fire disaster information, topographic information of each forest area, fire source information and other information.

3. Future prospects and implication of this system

The development and application of this system provides a new technical means for discovering possible forest fire hazards as soon as possible, and is of positive significance to reduce the fire impact on forest resources. The effective combination of image recognition technology, UAV technology, [5] infrared thermal imaging technology and the system will further improve the performance of the system.

4 Conclusion

Forests provide rich green wealth for human society, protect a large number of animal and plant resources, conserve water and soil, and maintain the self balance of the atmosphere on which human beings depend. To prevent and reduce the loss of forest fire as much as possible is not only to protect forest resources themselves, but also to protect human living environment and human beings themselves. Therefore, it is of great significance for the control of forest fire. During the research and development of the system, many specific problems to be solved in forest fire early warning and prediction have been encountered, and rich relevant knowledge and experience have been accumulated, which provides knowledge and experience accumulation for the subsequent research and development of relevant systems.

Acknowledgments

In the process of writing this article, the author has received the help and guidance of many experts and scholars. I would like to express my heartfelt thanks. In addition, a lot of network information has also been of great help to the author. Much thanks to them too. This study is supported by Shandong Agricultural Science and Technology Fund Project (2019ly004).

References

1. J. Huang, S. Lee, J. Zhang, et al. *J. Fujian agri. Fore. Univ*, **50**,1(2021).
2. R. Liu, H. Duan, X. Wu, *J.Anhui Agric.Sci.* **49**,4(2021)
3. H. Bai, D. Liu, S. Niu, et al. *J. Beij. Fore. Univ,* **43**,5(2021)
4. H. Hu, *Fore. fire ecology manag.* M(2005)
5. S. Liu, W. Yao, Y. Zhang, *Comp. Engi. Sci.* **42**,7(2020)

Study on the development of sansheng space in pastoral areas of Inner Mongolia

Huan Qu[1*], and *Jun* Ma[2]

Inner Mongolia University of Technology

Abstract. We should take the initiative to seize opportunities, scientifically respond to challenges, reasonably solve problems existing in development. effectively promote the accelerated transformation of development mode and the optimization and upgrading of industrial structure in pastoral areas of Inner Mongolia, and promote their coordinated development, innovative development, green development and modernization.By analyzing the development status of ecology, production and life in pastoral areas of Inner Mongolia, this paper finds out the problems existing in its development, and puts forward policy suggestions.
Keywords: Three living spaces; green development; Inner Mongolia Pastoral Area.

1 Current situation of green development in pastoral areas of Inner Mongolia

Green development in pastoral areas of Inner Mongolia has always been an important development issue concerned by the autonomous region and the whole country. This paper analyzes the current situation of green development in Inner Mongolia Pastoral Area from the ecological environment, industry and population of Inner Mongolia Pastoral Area.

1.1 Analysis of ecological status in pastoral areas of Inner Mongolia

1.1.1 Analysis on the current situation of grassland quantity and quality

The grassland in Inner Mongolia has been seriously degraded to an alarming extent. Among the available grassland area of more than 63.59 million hectares, the degraded grassland area has reached 38.67 million hectares, accounting for 60% of the available grassland. According to statistics, by the beginning of this century, the degraded area of grassland had reached 73% of the total grassland area, and the average livestock carrying capacity of grassland was 27.79 mu / sheep unit, down 38.03% compared with the 1980s. It is mainly reflected in the reduction of grass coverage and grass yield per unit area due to the influence of rats, pests and fire [1]. Rodent and insect pests and other bad soil layer structures and grass roots cause large-area grassland degradation and desertification, and destroy the ecological balance of grassland. It can be seen from the figure below that the overall trend of grassland rodent

*Corresponding author: 1039930437@qq.com

damage area and grassland fire area in pastoral areas in recent ten years is an upward trend, while the grassland pest damage area has gradually eased and is in a downward trend in recent years.

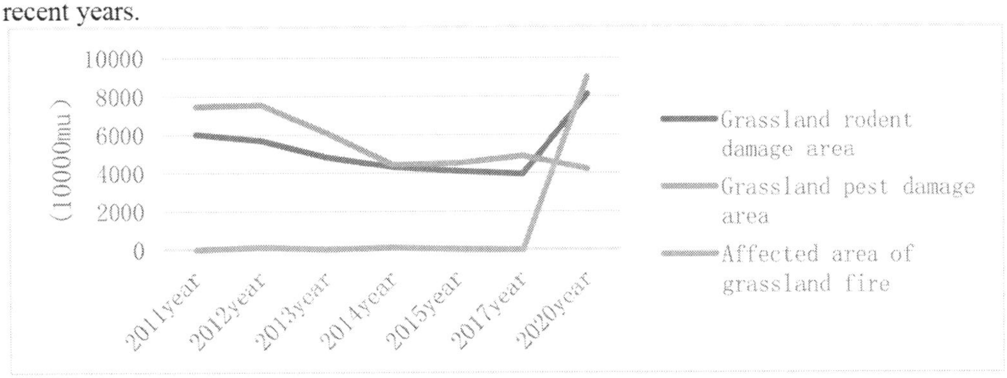

Note: data source Inner Mongolia statistical yearbook 2021 (missing data in 2016, 2018 and 2019)

Fig.1. Area affected by rats, insects and fire in Inner Mongolia Grassland.

1.1.2 Change of ecological assets

Ecological assets are natural assets that can provide ecological products and services for human beings. The quality of ecological assets is directly related to the supply and regulation of ecosystem and the change of cultural service value [2]. Through the calculation of ecological assets data of grassland and pastoral areas in Inner Mongolia in 2015, the area of grassland ecological assets is 436900 km2, accounting for about 57% of the total area of regional ecological assets; Through data analysis, taking 2000 as the dividing point, before 2000, due to land reclamation, the area of the whole region decreased by 3 million Hm2 every five years, and grassland and pastoral areas accounted for 50%; Since 2000, the grassland area in the whole region has hardly changed, and the grassland area in pastoral areas has decreased by 732.3 km2.

1.2 Current situation of industry in pastoral areas of Inner Mongolia

1.2.1 Changes in stocking capacity

The change of livestock carrying capacity reflects the production capacity of natural grassland in this area.Affected by policy support and rising prices, cattle and sheep breeding has obvious benefits, and the stock and sale of cattle have increased steadily.The slaughter of pigs and poultry increased.First, affected by drought, forage yield decreased significantly, and farmers increased their marketing to reduce costs;Second, with the implementation of the second round of grassland ecological reward and subsidy mechanism policy, all localities have increased the supervision and management of grassland livestock carrying capacity;Third, the price rise of pigs and poultry this year is also a main reason for accelerating the release of pigs and poultry.

Table 1. Number and growth rate of livestock in 2020.

index	2020year	Increase over the previous year (%)
Total meat output (10000 tons)	260.7	1.5
pork	61.4	-1.9
beef	66.3	3.9
mutton	113	2.9
poultry	20.1	-3.0
Total number of livestock at the end of the year (10000, PCs.)	7279.4	3.5
pig	534.1	24.3
cattle	671.1	7.2
sheep	6074.2	1.6

It can be seen from table 1 that the total meat output of Inner Mongolia Autonomous Region in 2020 was 2.607 million tons, an increase of 1.5% over 2019. Except for the decline of pig and poultry meat output, the other output and the total number of livestock showed an upward trend. Generally speaking, the improvement of Urbanization level increased the difficulty of reconciliation between traditional animal husbandry and modern production,It is not only reflected in the change of livestock carrying capacity, but also reflects the lack of characteristics of animal husbandry development management and weak ecological carrying capacity in pastoral areas, which has a certain impact on animal husbandry development.

1.2.2 Current situation of industrial and agricultural production

Agriculture and animal husbandry is the basic industry in Inner Mongolia Pastoral Area.According to the international measurement indicators of agricultural development level, developed countries account for 50% or more. According to relevant statistics, the highest value in China in previous years is 32%. Among them, the contribution of Inner Mongolia can reach 50%, and the proportion of pastoral areas is larger.The added value of agriculture, forestry, animal husbandry and fishery has been growing positively in recent ten years, reaching 167.771 billion yuan by 2017, the highest value over the years.The output of main crops basically kept rising.In recent 10 years, the secondary industry in pastoral areas has developed rapidly, including processing industry and mining industry. According to the data of the National Bureau of statistics, compared with 2019, most of the industrial output, such as crude oil, natural gas, coke and steel, has increased positively, and only a few of the industrial output is flat or reduced, such as the output of raw salt, soda ash and ethylene.The output of the mining industry also increased steadily from 2011 to 2016.

1.3 Population change in pastoral areas

The change of population in pastoral areas reflects the economic development level of pastoral areas to a certain extent. According to the data from Inner Mongolia statistical yearbook, comparing the data of pastoral population and urban population in 2019 and 2020, the urban population is 16.215 million. The rural population is 7.813 million; The urbanization rate of permanent residents reached 67.5%, with a birth rate of 8.3%; The mortality rate is 5.9% and the natural population growth rate is 2.5%. Generally speaking, the "grazing prohibition policy" has continuously improved the urbanization level of the permanent population in Inner Mongolia Autonomous Region, reduced the population in pastoral areas, and deepened the aging degree of herdsmen.

2 Problems of green development in pastoral areas of Inner Mongolia

2.1 The quantity and quality of grassland in pastoral areas of Inner Mongolia are decreasing, the grass and livestock are unbalanced, and the output value of animal husbandry needs to be improved

The problem of ecological balance between grassland and livestock is still significant. 70% of grassland in Inner Mongolia pastoral area exceeds the actual livestock carrying capacity. The allocation of agricultural and animal husbandry resources is unbalanced and lacks an effective combination model.According to the statistical yearbook, the total grassland area in Inner Mongolia is 78804.48 thousand hectares, and the grassland area is still decreasing in the past 20 years.The main reasons are: the development of urbanization, the development of industry and mining industry, etc;Affected by human and natural factors, the degraded area of grassland is increasing year by year, resulting in the decline of grassland quality.

2.2 Deterioration of ecological environment and decline of ecological function in grassland and pastoral areas

Agriculture and animal husbandry is the basic industry in pastoral areas of Inner Mongolia.According to the international measurement indicators of agricultural development level, developed countries account for 50% or more. According to relevant statistics, the highest value in China in previous years is 32%, which will drop to 29% in 2018. Among them, the contribution of Inner Mongolia can reach 50%, and the proportion of pastoral areas is larger . The added value of agriculture, forestry, animal husbandry and fishery has been growing positively in recent ten years. In recent 10 years, the secondary industry in pastoral areas has developed rapidly, including processing industry and mining industry. The output of the mining industry also increased steadily from 2011 to 2016.

2.3 Outflow of labor force and reduction of labor level

The unreasonable population structure in pastoral areas of Inner Mongolia, The education level and skill level of the population in pastoral areas are low,And the slow rate of income increase, affect the income of herdsmen, and the characteristics of population aging are obvious . Therefore, the level of education, health and management in some rural pastoral areas is also significantly declining, exposing the problems of low efficiency and high cost. With the development of science and technology , the vocational skills of rural pastoral workers can not keep up with the requirements of the modernization of grassland animal husbandry.

3 Suggestions on green development in pastoral areas

3.1 Optimal construction of ecological pastoral areas

Protecting and restoring ecology is the top priority in the construction of ecological pastoral areas, which mainly focuses on two aspects: first, optimizing and adjusting the balance between grass and livestock, and evaluating the production and ecological status of grassland through grassland dynamic monitoring; The second is Grassland quality and important ecological space protection. It is of great significance to delimit and strictly abide by the red line of ecological protection, promote natural resource protection and ecological restoration, and optimize the pattern of land and space development.

3.2 Development of animal husbandry and rational exploitation of mineral resources

Inner Mongolia Autonomous Region is rich in mineral resources .According to the relevant arrangements of the central government, since 2016, the autonomous region has successively carried out special actions such as implementing the rectification of the central environmental protection inspector and the special rectification of violations and violations in the field of coal resources, and achieved positive results, but there are still some weaknesses .Main measures: strictly control new mineral resources development projects on the grassland;Strictly regulate the existing mineral resources development projects on the grassland;Strictly supervise the development of grassland mineral resources;We will severely crack down on the illegal occupation of grasslands in the development of mineral resources.

3.3 Attract local talents to return home and build a team of animal husbandry support talents

The construction of talent team has always been an important factor in the development of rural industry. Actively build a new pattern of talent cultivation and attraction policies, continue to expand the Rural Revitalization talent team, and actively give full play to the leading and exemplary role of local sages and squires in industrial development and rural governance, so as to make them become the backbone of promoting rural development and progress. At the same time, build a scientific and technological talent team with reasonable structure to support agriculture and animal husbandry, and improve the relevant management system and assessment system, so as to make it an important promoter of professional and information management in the process of industrial development in rural pastoral areas.

4 Conclusion

The green development of pastoral areas in Inner Mongolia is inseparable from the effective coordination between the three living spaces. It not only promotes the economic development of pastoral areas, but also plays an important role in ecological protection.

Acknowledgement

Fund Project: National Natural Science Foundation of China (No.: 71864026); Natural Science Foundation of Inner Mongolia Autonomous Region (No.: 2018ms07004)

References

1. Miao Jing, Gai Zhiyi Thoughts on the evolution of industrial structure of 33 typical animal husbandry banners (cities) in Inner Mongolia [J]Inner Mongolia Social Sciences (Chinese version), 2018,39 (03): 182-188
2. Zhang Hao Study on sustainable development of grassland animal husbandry in Inner Mongolia [D] Northwest University of agriculture and forestry science and technology, 2019 DOI:10.27409/d.cnki. gxbnu. 2019.000140.

Study on influencing factors and monitoring cycle of wall thickness at key locations of natural gas station

Wenchen Cao*, *Ying* Song, *Liang* Feng, and *Shiyu* Gao
PipeChina Southwest Pipeline Company, China

Abstract. In recent years, the safe and stable operation of natural gas pipeline has become the focus of the whole society. Long distance natural gas pipelines are mostly large-diameter and high pressure pipelines, so their safe operation is particularly critical. In the high-pressure natural gas pipeline transportation industry, the wall thickness of key points (elbows and tees) will be thinned due to long-term erosion of gas, resulting in serious consequences. Therefore, its wall thickness monitoring is a key work of long-distance natural gas pipeline operation units. At present, most of them adopt off-line monitoring, but the standardized monitoring cycle cannot be effectively determined, and there are no relevant standards in the industry. The author uses fluent simulation software to establish the model, and then modifies the model combined with the field data. Then, the model is used to analyze the influencing factors of key points (elbow and TEE) one by one, and select more important influencing factors to formulate the monitoring cycle of different points. The monitoring cycle can provide reference for the daily work of on-site operators, and then ensure the safe and stable operation of long-distance natural gas pipeline.
Keywords: Long distance natural gas pipeline; Influencing factors; Wall thickness; Monitoring cycle; Elbow; Tee.

1 Introduction

Pipeline wall thickness detection is common in oil and gas field gathering and transmission pipelines and natural gas pipelines with high water content. For the long-distance natural gas pipeline, because most of the natural gas transported is purified and there is a filter separator for water removal in each station, few people at home and abroad have studied its wall thickness, and there is also a lack of relevant standards for its wall thickness detection. Most long-distance natural gas pipelines are large-diameter and high pressure pipelines, which will have serious consequences in case of leakage. Therefore, it is necessary to study the wall thickness of key points of long-distance natural gas station.

Throughout the relevant references, the research mainly focuses on the elbow of the pipeline. Yang Decheng et al. studied the erosion law and maximum erosion position of solid particles on the lower elbow. The results show that the particle velocity has an exponential relationship with the erosion rate; The increase of pipe diameter will reduce the

* Corresponding author: caowc@pipechina.com.cn

erosion rate of elbow, and the two show an exponential relationship; The particle diameter basically does not affect the maximum erosion position [1]; Song Xiaoqin and others studied the erosion wear failure of 90 ° elbow in gas transmission pipeline according to the actual working conditions on site. The results show that the greater the gathering pressure is, the less the erosion wear degree of elbow is; Gravity will affect the erosion wear degree and wear area of elbow; The smaller the gathering pipe diameter and the larger the particle diameter, the more serious the erosion wear [2]; Ji Chuling et al. Simulated the erosion wear of smooth inner wall and three bionic inner wall bends: transverse groove, longitudinal groove and pit. The results show that under the same conditions, the wear resistance of the thickened elbow can be significantly improved by using the bionic surface, and the wear resistance of the thickened elbow with different surface morphology is in the order of transverse groove shape, longitudinal groove shape, pit shape and smooth shape from large to small [3]; Zheng Yunping and others used fluent to simulate the relationship between various influencing factors and the erosion position and degree of elbow, and came to the conclusion that gas erodes the inner wall of elbow in a large area, and metal chips erode locally in this area; According to the different erosion conditions of elbows at different angles, the protection idea of "fixed side, large area and emphasis" is put forward in terms of protection parts and effectiveness [4]; Hulin deeply studied the wear law and on-line monitoring of wall thickness of fluid transmission pipeline with high temperature, high pressure and high solid holdup. The results show that the wear area and wear amount of elbow are quite different with different bend diameter ratio; At the same pipe diameter, the maximum wear decreases with the increase of bending diameter ratio [5].

The author uses fluent simulation software to establish the model, and then modifies the model combined with the field data. Then, the model is used to analyze the influencing factors of key points (elbow and TEE) one by one, and select more important influencing factors to formulate the monitoring cycle of different points. The monitoring cycle can provide reference for the daily work of on-site operators, and then ensure the safe and stable operation of long-distance natural gas pipeline.

2 Introduction to the built model

This paper uses the desinmodeler module of ANSYS Workbench software to draw the geometric model, the drawn elbow model and tee model.

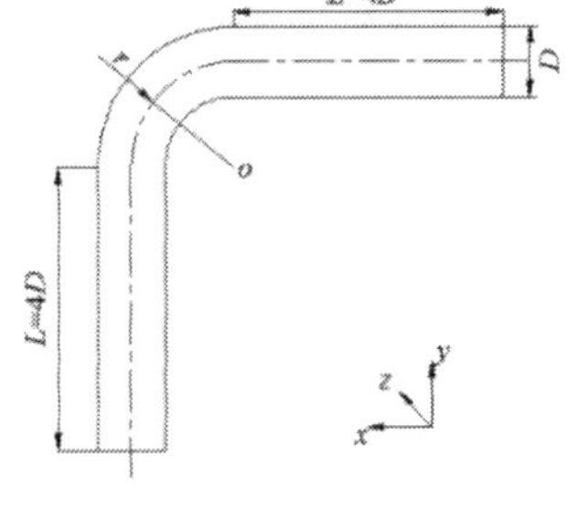

Fig. 1. Schematic diagram of elbow structure.

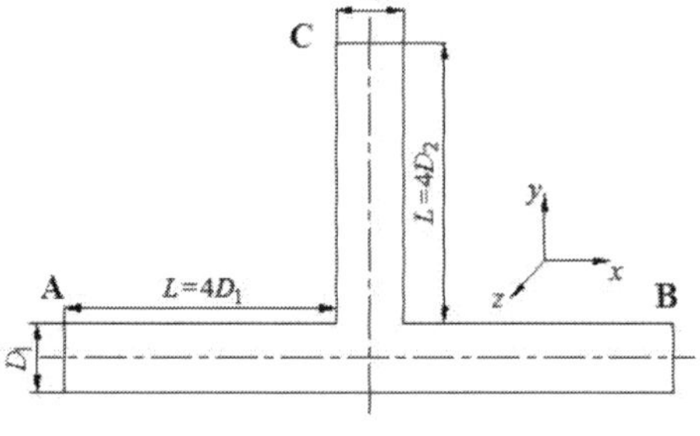

Fig. 2. Schematic diagram of tee structure.

Through the mesh module of ANSYS Workbench software, the two types of simulation models are meshed by using the proximity and curvature size function, and an expansion layer is set at the boundary to control the average quality of the grid.

During the simulation, the inlet and outlet boundary conditions are set as pressure inlet and pressure outlet. The relevant boundary conditions of fluid are set according to the parameters of specific working conditions. The gas phase material is methane, and the methane density under standard conditions is ρ= 0.714kg/m^3, μ= one point zero eight seven

× 10^{-5}kg/(m•s). Based on the field test data feedback model, the sand particle size is 0.4mm, which is close to the actual data. The particle size of the sand was used in the ancient study.

3 Research on influencing factors

3.1 Influence of fluid velocity on erosion

In the established erosion simulation model, the sand mass flow is set as 1 × 10^{-3}kg/s, and set the fluid velocity at the inlet of the model as 5m/s, 7m/s, 9m/s, 11m/s and 13m/s respectively. The results are shown in Figure 3.

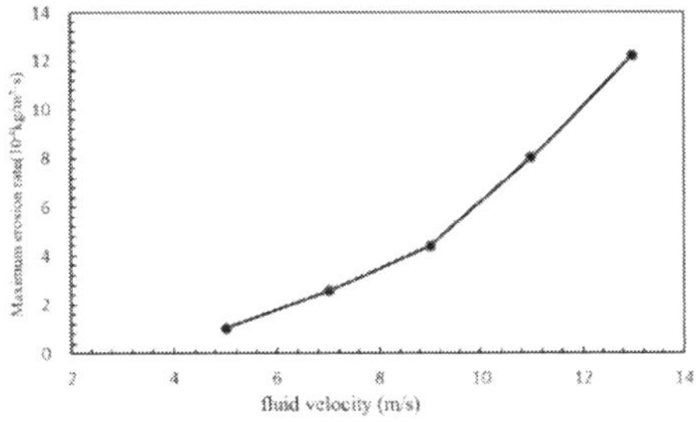

Fig. 3. Curve of fluid velocity and maximum erosion rate.

It can be seen that the maximum erosion rate at the most serious erosion point, that is, the monitoring point, increases with the increase of fluid flow rate. When the flow velocity is small, the maximum erosion rate increases slowly. When the flow velocity is large, the maximum erosion rate increases rapidly, showing an approximate exponential growth.

3.2 Influence of pipe diameter on erosion

In the established erosion simulation model, the mass flow rate is set as 0.02kg/s, the inlet fluid flow rate is set as 9m/s, and the elbow diameter of the geometric model is set as 150mm, 200mm, 250mm, 300mm, 400mm, 500mm and 900mm in turn. Other parameters of the model remain unchanged. The relationship between the results obtained after simulation calculation and the maximum erosion rate is shown in Figure 4.

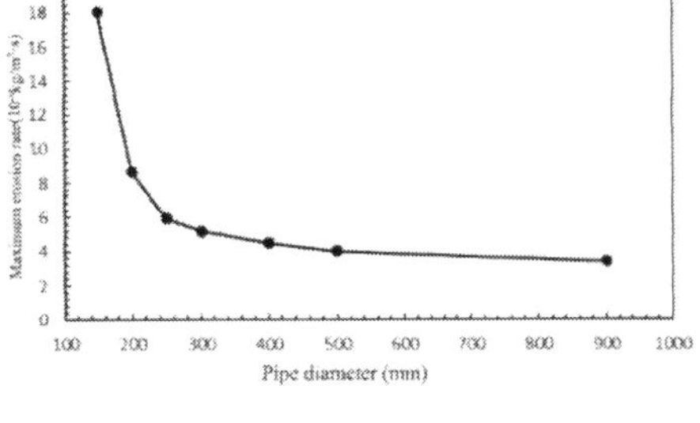

Fig. 4. Curve of pipe diameter and maximum erosion rate.

It can be seen from the figure that the erosion rate decreases rapidly with the increase of pipe diameter, and when the pipe diameter increases from 150mm to 400mm, the erosion rate increases from 1.825 × 10^{-5}kg/m^2•s reduced to 4.511 × 10^{-6}kg/m^2•s, which varies greatly. When the pipe diameter exceeds 400mm, the erosion rate is at a relatively low level. With the continuous increase of pipe diameter, the erosion rate decreases very slowly.

3.3 Influence law of sand mass flow on erosion

In the established simulation erosion model, keep the inlet velocity of 9m/s unchanged, set the sand mass flow to 0.02kg/s, 0.04kg/s, 0.06kg/s, 0.08kg/s and 0.1kg/s step by step, and keep other parameters unchanged. Conduct numerical simulation on pipe models of different sizes respectively. The relationship between the maximum erosion rate and sand mass flow is shown in Figure 5.

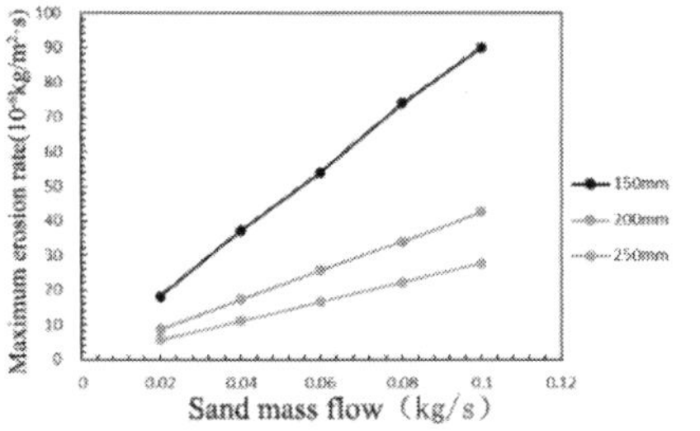

Fig. 5. Curve of sand mass flow and maximum erosion rate.

The relationship between sand mass flow and erosion rate is shown in the figure. It can be seen that with the increase of sand mass flow, the erosion rate also increases linearly. This is because with the increase of sand flow, the sand content per unit time increases and the number of collisions with pipe wall increases, resulting in the increase of pipe wall wear. It can also be concluded that with the increase of pipe diameter, the increasing speed of erosion rate with the increase of sand flow decreases gradually. Therefore, when the pipe diameter is small, the influence of sand flow on erosion rate is particularly obvious.

3.4 Influence of bend guide on erosion

When establishing the elbow simulation model, according to the actual installation mode of the elbow in Guangyuan gas transmission station, the simulation models of the elbow guidance as horizontal inlet, vertical upward inlet and vertical downward inlet are established, in which the first group of models are horizontal inlet and vertical downward outlet; The second group of models are vertical downward inlet and horizontal outlet; The third group of models are horizontal inlet and vertical upward outlet; The fourth group of models are vertical upward inlet and horizontal outlet; The inlet and outlet of the fifth group of models are horizontal. Other parameter settings remain unchanged. The simulation results are shown in Figure 6 and Figure 7.

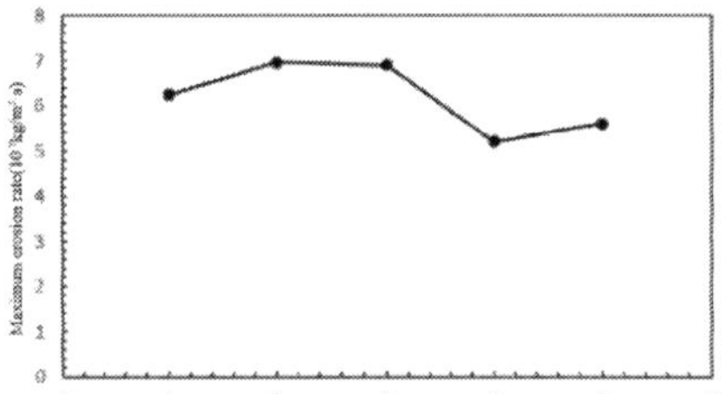

Fig. 6. Model and maximum erosion rate curve of different elbow guidance.

Figure 6 shows the impact of the inlet and outlet guidance of the elbow on the erosion rate. It can be found that the change of the erosion rate is not large. The erosion rate of the elbow is the largest under the conditions of vertical downward inlet and horizontal outlet, and the erosion rate of the elbow is the smallest under the conditions of vertical upward inlet and horizontal outlet. This is due to the increase or decrease of the incident velocity of

https://doi.org/10.1051/e3sconf/202234101029

sand particles under the action of gravity acceleration.

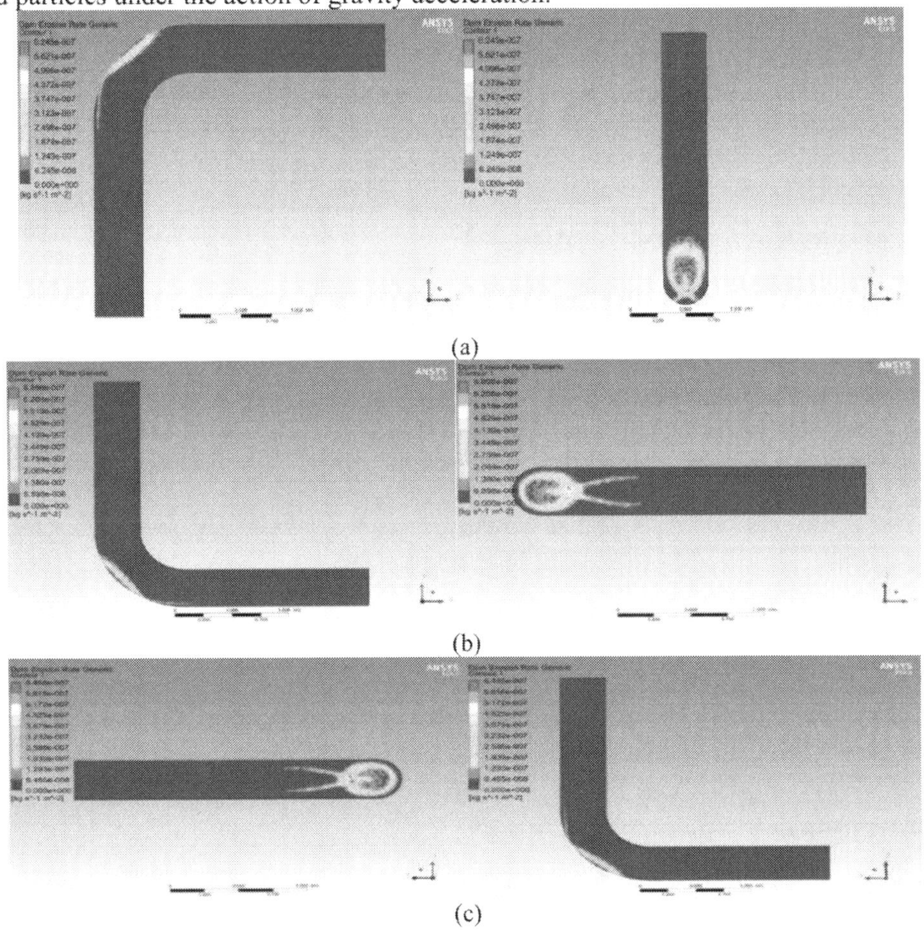

Fig. 7. (a) Elbow erosion at horizontal inlet and vertical downward outlet; **(b)** Elbow erosion at vertical downward inlet and horizontal outlet; **(c)** Elbow erosion at horizontal inlet and outlet.

It can be seen from (a-c) in Figure 7 that the elbow guide has no obvious influence on the erosion morphology, and the maximum erosion points on the elbow appear at the elbow outlet.

3.5 Influence of bend diameter ratio on erosion

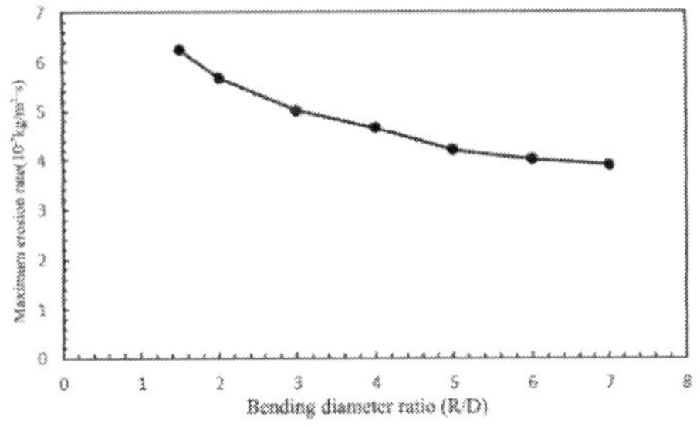

Fig. 8. Bending diameter ratio and maximum erosion rate.

When establishing the erosion simulation model, keep the inlet velocity 9m/s unchanged, change the bending diameter ratio (R/D) of the geometric model to 1.5, 2, 3, 4, 5, 6 and 7 respectively, and the settings of other boundary conditions remain unchanged. The results obtained after numerical simulation are shown in Figure 8.

Figure 8 shows the influence law of the bend diameter ratio of the elbow on the erosion rate. The erosion rate increases with the increase of the bend diameter ratio, and the decreasing speed of the erosion rate will slow down with the increase of the bend diameter ratio.

4 Determination of test cycle

According to Q / SY 93-2007 natural gas pipeline inspection specification, the general inspection of natural gas pipeline is once a year, and the first inspection shall be carried out within half a year after putting into use. The professional inspection covering pipe wall corrosion inspection shall not exceed 8 years because it is determined according to the actual safety situation of the pipeline. The professional inspection of new pipeline shall be carried out within 3 years after putting into operation, After 3 years, the inspection cycle shall be determined according to the pipeline safety status.

In the natural gas station, combined with the research results of the previous chapter, the following table 1 is divided by pipe diameter and pressure.

Table 1. Division of inspection cycle of station pipeline.

	less than 4MPa	less than 6Mpa	less than 10Mpa
less than 4Mpa 300mm	Once every 4 years	Once every 3years	Once every 1 year
less than 4Mpa 600mm	Once every 3 years	Once every 2years	Once every 1 year
Not less than 600mm	Once every 2 years	Once every 1 year	Once every 1 year

5 Conclusion

1. When the pipe diameter is small (below 400), the flow velocity, pipe diameter, sand flow and curvature are the main influencing factors; When the pipe diameter is larger (more than 400), the flow velocity and curvature are the main influencing factors.

2. The recommended cycle is shown in the table.

References

1. Yang Decheng, Zhu Hongwu. Erosion wear analysis of natural gas sand carrying gas-solid two-phase flow at elbow [J]. Petroleum machinery, 2019, 047 (010): 125-132
2. Song Xiaoqin, Liu Ling, Luo Songyang, et al. Study on erosion wear law of 90 ° elbow of natural gas gathering and transmission pipeline [J]. Lubrication and sealing, 2018, 043 (008): 62-68
3. Ji Chuling. Numerical simulation of bionic wear resistance method of elbow [J]. Journal of Central South University: Natural Science Edition, 2016 (47): 3589
4. Zheng Yunping, Wang Huanhuan, Yi Haolin, et al. Simulation study on erosion and protection of natural gas pipeline elbow [J]. Computer simulation, 2015, 32 (8): 427-430
5. Fu Lin. impact wear prediction and wall thickness monitoring of elbow of oil coal slurry transmission pipeline [D]. Hebei University of technology, 2009

AUTHOR INDEX